SMART CITY Barcelona

The Catalan Quest to Improve Future Urban Living

The Cañada Blanch / Sussex Academic Studies on Contemporary Spain

General Editor: Professor Paul Preston, London School of Economics

A list of all published titles in the series is available on the Press website. More recently published works are presented below.

Peter Anderson, *Friend or Foe?: Occupation, Collaboration and Selective Violence in the Spanish Civil War.*

Margaret Joan Anstee, *JB – An Unlikely Spanish Don: The Life and Times of Professor John Brande Trend.*

Richard Barker, *Skeletons in the Closet, Skeletons in the Ground: Repression, Victimization and Humiliation in a Small Andalusian Town – The Human Consequences of the Spanish Civil War.*

Germà Bel, *Infrastructure and the Political Economy of Nation Building in Spain, 1720–2010.*

Germà Bel, *Disdain, Distrust, and Dissolution: The Surge of Support for Independence in Catalonia.*

Carl-Henrik Bjerström, *Josep Renau and the Politics of Culture in Republican Spain, 1931–1939: Re-imagining the Nation.*

Kathryn Crameri, *'Goodbye, Spain?': The Question of Independence for Catalonia.*

Pol Dalmau, *Press, Politics and National Identities in Catalonia: The Transformation of La Vanguardia, 1881–1931.*

Mark Derby, *Petals and Bullets: Dorothy Morris – A New Zealand Nurse in the Spanish Civil War.*

Francisco Espinosa-Maestre, *Shoot the Messenger?: Spanish Democracy and the Crimes of Francoism – From the Pact of Silence to the Trial of Baltasar Garzón.*

María Jesús González, *Raymond Carr: The Curiosity of the Fox.*

Helen Graham, *The War and its Shadow: Spain's Civil War in Europe's Long Twentieth Century.*

Angela Jackson, *'For us it was Heaven': The Passion, Grief and Fortitude of Patience Darton – From the Spanish Civil War to Mao's China.*

Gabriel Jackson, *Juan Negrín: Physiologist, Socialist, and Spanish Republican War Leader.*

Nathan Jones, *The Adoption of a Pro-US Foreign Policy by Spain and the United Kingdom: José María Aznar and Tony Blair's Personal Motivations and their Global Impact.*

Xavier Moreno Juliá, *The Blue Division: Spanish Blood in Russia, 1941–1945.*

David Lethbridge, *Norman Bethune in Spain: Commitment, Crisis, and Conspiracy.*

Antonio Miguez Macho, *The Genocidal Genealogy of Francoism: Violence, Memory and Impunity.*

Carles Manera, *The Great Recession: A Subversive View.*

Nicholas Manganas, *Las dos Españas: Terror and Crisis in Contemporary Spain.*

Jorge Marco, *Guerrilleros and Neighbours in Arms: Identities and Cultures of Antifascist Resistance in Spain.*

Emily Mason, *Democracy, Deeds and Dilemmas: Support for the Spanish Republic within British Civil Society, 1936–1939.*

Soledad Fox Maura, *Jorge Semprún: The Spaniard who Survived the Nazis and Conquered Paris.*

Martin Minchom, *Spain's Martyred Cities: From the Battle of Madrid to Picasso's* Guernica.

Olivia Muñoz-Rojas, *Ashes and Granite: Destruction and Reconstruction in the Spanish Civil War and Its Aftermath.*

Linda Palfreeman, *Spain Bleeds: The Development of Battlefield Blood Transfusion during the Civil War.*

Isabelle Rohr, *The Spanish Right and the Jews, 1898–1945: Antisemitism and Opportunism.*

Rúben Serém, *Conspiracy, Coup d'état and Civil War in Seville, 1936–1939: History and Myth in Francoist Spain.*

Gareth Stockey, *Gibraltar: "A Dagger in the Spine of Spain?"*

Maria Thomas, *The Faith and the Fury: Popular Anticlerical Violence and Iconoclasm in Spain, 1931–1936.*

Dacia Viejo-Rose, *Reconstructing Spain: Cultural Heritage and Memory after Civil War.*

Antoni Vives, *SMART City Barcelona: The Catalan Quest to Improve Future Urban Living.*

Antoni Vives

SMART CITY Barcelona

The Catalan Quest to Improve Future Urban Living

Translated from the Spanish by David Thomas Clark

sussex
ACADEMIC
PRESS
Brighton • Portland • Toronto

Cañada Blanch Centre
for Contemporary
Spanish Studies

CATALAN
OBSER
VATORY
LSE

2 4 6 8 10 9 7 5 3 1

First published in Spanish: *SMART. Las ideas que convirtieron a Barcelona en una ciudad líder en el mundo*, by Arpa y Alfil Editores, S.L., 2017.

First published in Great Britain in 2018 by
SUSSEX ACADEMIC PRESS
PO Box 139, Eastbourne BN24 9BP

Distributed in North America by
SUSSEX ACADEMIC PRESS
ISBS Publisher Services
920 NE 58th Ave #300, Portland, OR 97213, USA

Published in collaboration with the
Cañada Blanch Centre for Contemporary Spanish Studies and
the Catalan Observatory, London School of Economics.

British Library Cataloguing in Publication Data
A CIP catalogue record for this book is available from the British Library.

Library of Congress Cataloging-in-Publication Data
Names: Vives, Antoni, 1965– author.
Title: Smart city Barcelona : the Catalan quest to improve future urban living / Antoni Vives.
Other titles: Smart. English
Description: Brighton ; Portland : Sussex Academic Press, [2018] | Series: The Cañada Blanch/Sussex Academic studies on contemporary Spain | "First published in Spanish: SMART. Las ideas que convirtieron a Barcelona en una ciudad lider en el mundo, by Arpa y Alfil Editores, S.L., 2017."
Identifiers: LCCN 2017044637 | ISBN 9781845199180 (pbk : alk. paper)
Subjects: LCSH: City planning—Spain—Barcelona. | Technological innovations—Spain—Barcelona. | Urban policy—Spain—Barcelona. | Barcelona (Spain)—Economic policy. | Barcelona (Spain)—Politics and government—21st century.
Classification: LCC HT169.S652 B38813 2018 | DDC 307.1/2160946712—dc23
LC record available at https://lccn.loc.gov/2017044637

MIX
Paper from responsible sources
FSC www.fsc.org FSC® C013056

Typeset & designed by Sussex Academic Press, Brighton & Eastbourne.
Printed by TJ International, Padstow, Cornwall.

Contents

The Cañada Blanch Centre for Contemporary Spanish Studies

In the 1960s, the most important initiative in the cultural and academic relations between Spain and the United Kingdom was launched by a Valencian fruit importer in London. The creation by Vicente Cañada Blanch of the Anglo-Spanish Cultural Foundation has subsequently benefited large numbers of Spanish and British scholars at various levels. Thanks to the generosity of Vicente Cañada Blanch, thousands of Spanish schoolchildren have been educated at the secondary school in West London that bears his name. At the same time, many British and Spanish university students have benefited from the exchange scholarships which fostered cultural and scientific exchanges between the two countries. Some of the most important historical, artistic and literary work on Spanish topics to be produced in Great Britain was initially made possible by Cañada Blanch scholarships.

Vicente Cañada Blanch was, by inclination, a conservative. When his Foundation was created, the Franco regime was still in the plenitude of its power. Nevertheless, the keynote of the Foundation's activities was always a complete open-mindedness on political issues. This was reflected in the diversity of research projects supported by the Foundation, many of which, in Francoist Spain, would have been regarded as subversive. When the Dictator died, Don Vicente was in his seventy-fifth year. In the two decades following the death of the Dictator, although apparently indestructible, Don Vicente was obliged to husband his energies. Increasingly, the work of the Foundation was carried forward by Miguel Dols whose tireless and imaginative work in London was matched in Spain by that of José María Coll Comín. They were united in the Foundation's spirit of open-minded commitment to fostering research of high quality in pursuit of better Anglo-Spanish cultural relations. Throughout the 1990s, thanks to them, the role of the Foundation grew considerably.

In 1994, in collaboration with the London School of Economics, the Foundation established the Príncipe de Asturias Chair of Contemporary Spanish History and the Cañada Blanch Centre for Contemporary Spanish Studies. It is the particular task of the Cañada Blanch Centre for Contemporary Spanish Studies to promote the understanding of twentieth-century Spain through research and teaching of contemporary Spanish history, politics, economy, sociology and culture. The Centre possesses a valuable library and archival centre for specialists in contemporary Spain. This work is carried on through the publications of the doctoral and post-doctoral researchers at the Centre itself and through the many seminars and lectures held at the London School of Economics. While the seminars are the province of the researchers, the lecture cycles have been the forum in which Spanish politicians have been able to address audiences in the United Kingdom.

Since 1998, the Cañada Blanch Centre has published a substantial number of books in collaboration with several different publishers on the subject of contemporary Spanish history and politics. An extremely fruitful partnership with Sussex Academic Press began in 2004. Full details and descriptions of the published works can be found on the Press website.

A constant interest of the series has been the historical roots of the present relationship between Catalonia and the political establishment in Madrid. In 2010, there was the study of the economic future of Catalonia and of the role being played in that future by the region's ports by Ramon Tremosa i Balcells, *Catalonia, An Emerging Economy*. In 2011, Olivia Muñoz-Rojas cast startling light on post-war reconstruction in *Ashes and Granite*. In 2012, Germà Bel's startlingly original *Infrastructure and the Political Economy of Nation Building in Spain, 1720–2010* exposed the damage done to the Catalan and also the Spanish economies by the country's asymmetrical and dysfunctional transport and communications model. In 2013, Andrew Dowling's *Catalonia since the Spanish Civil War: Reconstructing the Nation* examined the reconstruction of national consciousness in Catalonia since the Civil War. In 2014, in the run-up to the referendum on Catalan independence, bitterly opposed by Madrid, Kathryn Crameri's *GOODBYE, SPAIN? The Question of Independence for Catalonia* provided a key contribution to the debate. In 2015, in his *Disdain, Distrust and Dissolution: The Surge of Support for Independence in Catalonia*, Germà Bel returned to the series with a dissection of the problems of the relationship between Catalonia and Spain: the dynamics of conflict between groups, their effects on the inter-territorial distrust, and their impact on the functioning of the state.

All these have been important works about the past and the present of Catalonia. The present volume says much about both but crucially adds a vision for the future, Antoni Vives' *SMART CITY Barcelona: The Catalan Quest to Improve Future Urban Living* is an inspiring beacon of hope for how life in our clogged and faceless cities can use digital technology to accommodate the needs of communities and individuals by improving all kinds of services ranging from public transport and recreational space to waste collection and recycling.

Preface by Series Editor Paul Preston

This book is as elegantly and vividly written as one would expect from one of the most talented novelists writing in Catalan today. It is as empirically sound as one would expect from someone who spent several years as deputy mayor of Barcelona with responsibility for urban planning and several more incessantly travelling the world acquiring an enviable knowledge of the major cities in Latin America, the United States, Russia, China, India, the Middle East, Africa and most of Europe. Both in politics and in private practice, Antoni Vives has devoted himself with a passionate commitment to dealing with the central questions of urbanisation in the twenty-first century. He sees how the need to make cities easier places to live – whether in terms of transport, housing, health care, recreational spaces and the key issue that underlies all others, funding – is a social as much as a logistical problem. Toni has campaigned against the reduction of cities to amorphous conglomerations in which citizens are merely cogs in the overall economy. Instead, building on his own experience in Barcelona, he has recognized the central importance of maintaining, but also connecting, small neighbourhoods with a real sense of community.

In Barcelona and now beyond, he has been an evangelist of how to improve the lives of all citizens through the deployment of digital technology. What he calls the Social Smart City is an exciting concept of how that technology can improve all kinds of services ranging from public transport to waste collection and recycling. His knowledge of the problems facing, and the solutions awaiting, the world's clogged cities is unparalleled; it is based on his own practical experience and demonstrated in his many inspiring lectures at the London School of Economics.

I find this book exciting as I try to apply Toni Vives' vision to the cities of my own very limited experience. The cities I know best are Liverpool, where I was born and brought up; Madrid where I lived for four years from 1969 to 1973 and spent a lot of time in the subsequent years; London, where I have lived and worked since 1979; and Barcelona. I have never lived in the Catalan capital yet it is almost certainly the Spanish city that I have visited most times, at least fifty in the last forty years.

All four have changed dramatically in my lifetime and not always for the better. The Liverpool in which I grew up had been pulverized by the Luftwaffe during the Second World War. What the Luftwaffe started was finished by town councils. The streets of smog-blackened terraced houses that were characteristic of Liverpool's working-class areas were bulldozed in the 1980s and large areas of what was once housing are now empty areas, green but empty as the families that lived there were moved out to new towns on the outskirts. No sign there of the kind of enlightened vision of Toni Vives and his colleagues, none of the productive public space that they envisaged for Barcelona over the last twenty-odd years. I could say the same about the missing municipal vision for London, clogged by traffic, poisoned by pollution, torn by a housing crisis while much property belongs to foreign speculators and lies unoccupied.

The Madrid that I lived in was still very much like the Republican capital of the civil war years. There were streets that ran from the Puerta del Sol which contained large numbers of shops specialising in artificial limbs, their customers being largely soldiers and civilians mutilated during the Spanish Civil War. The house that I lived in between 1972 and 1973 was in the district of Argüelles near the University, an area that had been on the front line of the siege of the city by Franco's African columns. It had seen fierce fighting and bullet holes could still be seen in the walls of several surrounding streets. A year earlier, I had lived in a house in the old working-class district barrio of Tetuán and from the window of my room I looked on to barren fields where sheep grazed. Nearly half a century afterwards, that house is now many kilometres from the ring roads that now mark Madrid's outer boundaries. The city is bustling in terms of traffic and huge housing estates have grown up, but the sense of community of old Madrid has gone and there is no modern vision to replace it.

I lived in Madrid as part of the process crucial for any historian serious about researching another country's past, a process involving the mastery of a new language and culture, an immersion in the country which will at best lead to the acquisition of what can be called 'a second identity'. For most Anglo-Saxon historians of Spain that usually means becoming an adopted citizen of either Madrid or Barcelona. For me, Madrid was where I learned to speak Castilian and where I lived. Inevitably, that meant that when I went to Barcelona it was as a foreigner. The Barcelona that I first knew in the 1970s was hemmed in by the huge warehouses that cut off the port from the city. When I first went to Barcelona in the late 1960s, the strongest impressions were created neither by the Tibidabo nor by Montjuich, neither by the Sagrada Familia nor the Barri Gòtic. What I most remember is a chance meeting with a fifty-year-old veteran of the CNT (Confederación Nacional del Trabajo) who lived in a tiny ground-floor apartment in the drabbest part of the Paralelo. He invited me into his home where I shared a meal with his family. In the austere surroundings of his home, he reminisced about Barcelona during the Spanish Civil War and about the clandestine trade union struggles of the 1960s. Non-violent, rationalist, this was a man who was an authentic humanist, generous of spirit, full of solidarity for his neighbours and workmates, hopeful of rational solutions to the problems of humanity yet not naïve. It was an encounter which confirmed all my romantic, and probably wildly misplaced, illusions about workers' communities in Barcelona. This was not the Barcelona of Gaudí and Cambó but it gave me a link with the working-class Barcelona of the Republic and the Civil War.

That Barcelona no longer exists. The Olympic Games of 1992 opened up the city to the sea and change since has been vertiginous in terms both of the hordes of tourists and the slum clearances that have changed the Paralelo, the Ravall and Poble Nou. The changes have an overall coherence designed by the Catalan architect Oriol Bohigas at the Barcelona School of

Architecture and made possible by the drive of the Socialist mayor Pasqual Maragall. It is true that the SMART Barcelona envisaged by Toni Vives and digital visionaries like Manel Sanromà and Vicente Guallart is still a work in progress but it is far further along the way than Liverpool, London and Madrid and, no doubt, by many cities of which I know little.

Author's Acknowledgments

This book is the product of many hours of reflection, conversation and work. Above all, a lot of work. The following are the names of people that have urged me, encouraged me and helped me to write it, supporting me with generosity and good humour in the most difficult moments, making the easier moments far sweeter: in first place, my editor, Joaquim Palau, who from the day we signed the first contract on a napkin in the Paraigües bar in the Plaça de Sant Miquel de Barcelona, has never ceased to be one of my strongest and most critical supporters. Anthony Grahame, my British publisher, who right from the start has supported this adventure with enthusiasm and adding value to it. Dr. Jordi Amat, my young brother and teacher, without whose blessing I wouldn't dare write a single page. Jaume Roures, who always encourages me to be brave; this book forms part of his therapy. Oriol Soler, who's always there for me; an essential part of this book is his. Eduard Martín and Josep Ramon Ferrer, bravery, professionalism and intelligence: I owe some of the pages of this book completely to him. Joan Anton Sánchez, who brought more than a few cubic metres of wisdom from Fiésole; his generosity with me knows no bounds. Manel Sanromà, one of those older brothers that life gifts you, an International Brigadier of freedom and social justice, a knight-errant of intelligence, who also urged me to write *Smart*. My colleagues from IAAC, Vicente Guallart and Willy Müller, where we made everything start: this book belongs to this school too. Albert Civit, my right hand, my left hand and my all. His advice as a dear friend, without concessions, and his voice, are ever-present in every line. Abha Joshi-Ghani, who showed me the Indian way, for his prologue of which I am wholly undeserving; in fact, I think he may have confused the book and the author. Sulakshana Mahajan, who fights to bring Mumbai back to its citizens, and edits and translates this book to Maharati. Professor Paul Preston, teacher, friend and permanent inspiration for my daily work, this book included, of course: his mastering in dignity and life goes far beyond his mastering in history. Adam Austerfield, and my LSE friends who dared to believe in me. I owe him and all of them some of the best moments of my life, and the guts to put this text in English. Oscar Vela, with whom I share life and business management *clínics* while we walk along the Catalan mountains, recalling the best that basketball gave to us. There's much from that in this book. Joaquim Forn, my great friend in politics and in life, who tints everything with nobility and good heart: our Manentian way (after Albert Manent) of understanding public service and serving Catalonia, full of *Pacem in Terris*, is here. To Xavier Trias, the anti-populist mayor *par excellence*, mine and everyone's mayor: to him I owe being able to devote myself to what I do. My family, obviously: Ignasi, Maria, Toni, Elisabet. And Laura, who has withstood everything. Her steadfastness through the hardest times, her resilience, unconditional love and positive spirit inspire me daily. And finally, Maria Sisternas, my faithful friend: the Smart discovery of this part of my life. My accomplice and the driving force behind this book, and practically everything. The book is dedicated to her.

I would have liked to end these acknowledgements here, but I'm not able to. Albert Piedrabuena, a fine man, the person who was continuously at my side for so many years, working for Barcelona, left us a few months ago. His youth, his energy, his trust and his loyalty have been my fuel ever since we were the opposition. Afterwards came friendship and hours of walking together. I'm sure I will find him one of these days, halfway up Montserrat. I wrote this book for him.

Prologue

Abha Joshi-Ghani

Director for Knowledge Exchange and Learning at the World Bank Institute

At last Antoni Vives has decided to gift us his vast knowledge on cities in the form of a book. Antoni is one of the world's leading experts on citizenship, local governance, cities and the social constructs that define them. Antoni champions them as inclusive, fair, enterprising, Smart and with quality of life. He does so from his city, Barcelona, the city that many think has laid the foundations of what 21st century cities should aspire to be.

The book benefits from Antoni's direct experience in conducting the transformation and economic recovery of his city, whilst he was the Deputy Mayor of Urban Habitat, during one of the most difficult periods of global social and economic crisis in history. The book's virtue lies in analysing cases and cities from developed and developing countries in which he has worked, discovering global challenges common to all of them: inequality, unemployment, housing, the subversion of human spirit in the daily fight for life . . . As he likes to say, the challenge lies in the *urbanising* world, rather than in the developing world. The virtue of this book is that Antoni discovers paths to action to strengthen cities through concrete policies and actions.

Antoni Vives delves into the history of global cities like Barcelona and New York, and reveals the forces that make them what they are: architecture, city planning, social and cultural history, as well as sociopolitical structure. Antoni draws lessons that are immediately applicable to the fastest growing cities in the developing world, identifying the elements that make up this rapid and ever complex urbanisation: migration, poverty, a lack of decent housing, inequality, the search for a more dignified life, quality services and making all human aspirations attainable.

I dare say that in *SMART Barcelona*, Antoni offers us ten new lessons on the theory and the praxis of developing cities; how we can turn expanding cities, but also those that are subjected to the effects of the crisis, into Social Smart Cities, by making the public and private sectors, as well as civic forces that work for the inclusion of all humans, collaborate in the construction of a better, fairer society for better and wealthier lives for all. This book is a tour de force that distills his years of experience, as well as the passion and intellectual independence, that define him as a person . . .

I met Toni, that's what his friends call him, in 2011, when he had been recently named Deputy Mayor of Urban Habitat in the city of Barcelona. I was impressed by his and his team's obsession for the reconstruction of the city's civic and social apparatus. Barcelona had long been considered one of the world's most innovative cities, but institutions such as the World Bank had never before seen the capability that Toni demonstrated. His passion for the compact, dense, pretty (yes, aesthetics matter), vibrant, fair and social city stood out. Preserve the best of the old and embrace the best of the new, he told me smiling, is our way of understanding progress. What could we add to these ideas, strength? But Antoni Vives and his team contributed something else, and they elevated it to a different level, because its focus was on empowering communities, neighbourhoods, cities and how, using technology, to improve public services for citizens. Its focus, then, was twofold: they understood the city's challenges through the eyes of its directors and political leaders; but they also saw the city through the eyes of its most underprivileged residents, those who are left behind, those for whom we have to create new opportunities in more productive, inclusive and sustainable cities. Therefore we developed a very solid partnership between the World Bank and Barcelona, and as a result celebrated the Sixth World Bank Urban Symposium in the city, on the 12th of October 2012, with a title that was incredibly in line with Antoni's ideas: "Rethinking cities."

Toni's team's focus on neighbourhood improvement also defined his obsession to make cities socially sustainable. Toni sees cities as a web of interconnected and interrelated small neighbourhoods, formed of human bonds that make them inclusive and socially relevant. His concept of self-sufficiency surpasses the field of energy and becomes social: schools, markets, health centres, parks, shops, cafés, bookshops, libraries, museums, public transport . . . the city, therefore, is like that interconnected web of small communities and human relationships, that are facilitated but not guided by public entities.

But Toni was, and is, very demanding of himself. He isn't afraid to talk about his failures. Nor about their possible causes. "We can't have intelligent cities if their mayors are not intelligent," said Toni in the inaugural speech at the Smart City Expo and World Congress of 2013, in Barcelona. By this he meant that a Smart city must not be seen as a set of devices and sensors, nor a hectic search for the latest advances in technology. A city's Smart government, he said, is that which deploys urban services and transformations to better serve its citizens, using the available technology. Nothing more, nothing less.

In *SMART Barcelona* Antoni develops, through well-founded examples from the past, but also from his experiences in Barcelona and internationally, the idea of a Social Smart City, imploring city administrators and their experts to take advantage of the digital revolution and big data in order to provide better services for citizens: from traffic management and energy efficiency to zero emissions, waste management and recycling for energy generation; or the provision of improved social services, to name but a few. His emphasis is on the creation of robust policies based on flexible organisation, and on a governance that integrates the voice of citizens in the process of defining criteria, as well as their implication in the appropriation of services and public spaces, which is only possible through policies based on objective data.

In this book Anton Vives gifts us, as I was saying, his way of understanding and interpreting the city, how the city is shaped by its history, its art, and its architecture; by its spirit, by the *staatgeist* (state spirit) that resides in each and every one of its citizens. Being Smart is not being technological, he tells us. It's having the ability to define a vision. And how the future lies in making the most of the technological and digital revolution, to radically improve our cities and our democracies.

You are not in front of a book that's going to leave you feeling indifferent. Toni never leaves anyone feeling indifferent. Everyone that knows him, knows that. If it's the first time you've come across him, this book will confirm it. In fact, this book is a testament to his passion for cities, for human beings, for art and above all for literature. As a result, he infuses all of its pages with the unique character of his Renaissance-like thought.

SMART CITY Barcelona

The Catalan Quest to Improve Future Urban Living

1

What I've Learned

On the 24th of May 2015 we lost Spanish local and regional elections. We, who thought we'd found the right model. We, who exported the very same model all over the world, who had received prizes everywhere . . . we lost the elections, even if it was by a tiny margin. It was a big lesson in humility. Losing puts you in your place. We lost against a political option that had been put together very quickly, from the bottom up, allying itself to those who are used to being on the margins. Ada Colau, the new mayor, and her team gave us an authentic lesson in how to campaign in the modern age: position yourself on the sidelines of established politics; capitalise on the legitimacy of the just fight for the social rights of the most disadvantaged; give political communication a new tone, casual, authentic, liberated from the traces of the establishment that influenced our proposal, however strong it may have been. Keep pace with the pulse of the most popular neighbourhoods . . . Governing, though, is another thing . . . On the 24th of May 2015 I started to understand that I needed to unlearn a lot of what I had learned, that I would have to start again a lot of what I had started years ago. There wasn't space to mend what we had done. We had to start again, and that was exactly what quite a few of us did.

For the last two years I have explained in dozens of conferences that, among the mistakes we made, there are two that stand out: First of all, placing too much trust in the term "experience". Having experience matters little in a world that is changing daily. Furthermore, having experience moves you excessively close to the situation that already exists, to that which has been proven unable to give hope, to put an end to social inequality and lack of opportunity. Secondly, we weren't able to communicate effectively an essential idea: Barcelona's success is linked to its social bond, to that social dream of which I have spoken and of which I speak over and over again, wherever I go. The Barcelona that we relaunched revolved around the creation of opportunities for everyone. However this wasn't, and isn't, lived in this way by many sectors of our society. Success shouldn't be measured unless it is analysed in detail, if the only thing done is recording the usual indicators: whether GNP, household income, whether this index or the other . . . An inexcusable error. In the age of big data, in the age of digital information and of cross-analysis; in the age of algorithms applied to design and the provision of public services, applying judgment based on wild guesses isn't good enough in the 21st century. In the time of big data what really counts is small data.

Experience and a social dream. Hope. The incarnation of an inclusive, open, optimistic, possible, viable new truth . . . None of this was communicated well enough. We lost and we became the opposition. We went home. How it should be. After all, Barcelona has fought historically for democracy. We're proud of it. Barcelona is a city that likes to explain itself through its historical scars: the marks of the bombs in the Plaça de Sant Felip Neri; the bullets lodged forever in the Customs building in the Port of Barcelona; the secret police of Carrer de Sant Elies; the cells under the Vía Layetana; the anti-aircraft batteries of Turó de la Rovira . . . But also the large protests for freedom, for peace, for social rights, for Catalonia . . . Barcelona, this social dream, expresses itself democratically like few other cities in the world, and it does so with high-quality public services, public spaces, like few places on Earth.

The secret of my city is also rooted in the history of Catalanism: the sacred continuity and the social dream. Barcelona has been able to build from generation to generation through complementary visions, that have improved and updated it. I think that this is another of the great lessons of recent times. Barcelona picks up on the signs of the times, just as Paris, London, New York and Amsterdam have done, to name a few of the most important capitals. That's another great lesson: it's essential to properly read the times, above all in those human communities in which collectivism, matters. And collectivism, in Barcelona, is what's most important.

2015's lesson in humility led me to deep reflection, and a radical decision: I had to leave it all behind in order to start again. And that's what I did. I had relied on the best team that a person can have: the Urban Habit team in Barcelona City Council, people with a level of professionalism and a loyalty to their city that was unshakeable. I had to start again, to reflect on what I had learned with them, to create a renewed proposal, not only for Barcelona but for all the cities in the world. I started to work by travelling to the four corners of the Earth, helping cities and organisations as much as I could. This has allowed me, in the last two years, to contemplate our experience from the outside, to see what was really useable and discard what doesn't make sense to continue.

The most important thing I learned has to do with the scale of our task: cities are on a human scale, otherwise they don't make sense. A city is Smart, intelligent, or whatever it may be, simply if it has an authentic and possible social ambition. I'll give you an example to make it clear: today we can see how cities are consumed by these horrible *verbrauchentempels* that offer nothing more than urban emptiness. It's true that these soulless temples of consumption, built around the same old franchises, tend to have commercial success. The problem is that today we know that they have an expiry date, and it is the subcontracted companies themselves that realise this. What has happened at Ground Zero in New York's Lower Manhattan district is significant. Calatrava's transportation hub, aside from the debate on its aesthetic, is nothing compared to Grand Central's capacity and role. Whilst 50,000 people pass through the former daily, 750,000 pass through the latter. For the hub designed by Calatrava commerciality was essential. What an error! Today's great lesson is that there's no better shopping centre than an active street, the ground floors of our buildings up and running, with the blinds open. Grand Central has a few restaurants, a few shops . . . but above all it gives life to central Manhattan where life is more important: on the streets. Calatrava's vastly expensive totem not only doesn't do this, but in fact could put an end to the few commercial possibilities, and by extension social and civic possibilities, of its surroundings.

That is why rereading the city, unlearning, admitting mistakes, taking a closer look at successes and learning again is what has driven me to work on shaping a Smart proposal for cities across the world, during the last few years. This proposal has to do with the configuration of a social and operational vision, in the sense that it can be implemented, that it produces opportunities, that it activates the city; that it provides it with the elements that make it a space for social justice and equality of opportunity for those who dare to generate wealth, to generate employment and future.

I owe everything that I've learned to the people that have worked with me all these years. All of them will make an appearance in the pages that follow. I wasn't and am not anything without them. I'm nothing without the architectural, urbanistic, social and cultural tradition of my city: from Oriol Bohigas to Pasqual Maragall and up to Xavier Trias, to Albert Serratosa or Joan Antoni Solans (Spanish architects and politicians), who have complemented each other through their ideological postulates. I'm also nothing without the experience I have acquired working from China, to India, the Middle East, Africa, Europe and America. I'm not naming these countries for the sake of it: in each of these places I've had and have the opportunity to work on a daily basis with public and private servants that contribute freedom and ability to the international cities movement. Barcelona is great, of course, but it is dwarfed by the courage, bravery and intelligence of cities we have hardly heard of, if we've even heard of them at all.

That's another lesson in humility. Because what's European, what's Western, is great, but not so much so when looked at in light of history; in light of what we are setting out for the next two generations. Europe, the West, must prevail as a place for individual and collective liberty, a place for duties and social rights. Thus, what we need is constructive dialogue with social and cultural environments from which we have a lot to learn. If we do not do so, Europe is condemned to a slow but unstoppable decline. And, what do you want me to tell you, neither I, nor many of the people I have worked and work with, are going to resign ourselves to this fact.

All of this is what I have learned before and after the 24th of May 2015's lesson in humility. All of this and what follows in the pages of this book, written to reflect with everyone on the meaning of cities, on their present and future as spaces for the creation of presents, futures, hopes, freedoms, opportunities and well-being for as many people as possible.

The text to follow refers to the geography and districts of Barcelona. The official and most detailed map of the city, in English, is available at: https://w33.bcn.cat/planolBCN/en/

2

Why Smart

A few weeks after finishing the draft of this book, the majority of space in the Catalan media was occupied by one piece of news: an elderly woman had died in the city of Reus, about seventy miles south of Barcelona, when she accidentally set fire to her home. The woman, who couldn't afford to pay her gas bill, cooked and kept warm with candles. Of course, the inevitable happened: the home burnt down and the blaze killed the poor woman. The events in Reus demand explanations. For an elderly person, alone, poor and seemingly abandoned by society, to die by setting fire to her home whilst trying to keep warm with candles is incredibly serious, it's a collective failure. I got ready to listen to statements from politicians and Gas Natural[1] and it seemed to me that, other than trying to establish their stances towards the insurers and the fear of what they might say, there wasn't a clear acceptance of shared responsibility. Mònica Terribas, in one of her consistently good interviews that go out shortly after the morning programme on Catalunya Radio, cornered the managing director of Gas Natural when he clung to regulation, or to the lack of it, in order to excuse himself for not knowing how to take action. Neither does it seem that the government, nor the surrounding community, were aware of the errors. Or of the technology and the available data. Data and people. I am alluding to the network that we should all have, to be able to get by, to be able to live with dignity, feeling protected and loved: family, acquaintances, the council, the government. In fact, later on we found out that there was a daughter, relatives, neighbours . . .

Every November the Smart City Expo and World Congress takes place in Barcelona. For Manuel Sanroma, Managing Director of the

---

1 A Spanish natural gas company that has its headquarters in Barcelona.

Open Administration Consortium of Catalonia (AOC), it was clear, it wasn't enough just to hold the congress. We needed to work on its legacy and we got down to business: we put municipal machinery to the job. There's no dispute about it: Barcelona is the world capital of the unstoppable movement of Smart Cities, one of the driving forces behind the creation of jobs and some of the most important opportunities in the world. President Obama said, a few days after celebrating the victory of his final term (November 2012), that anyone interested in the improvement of cities in the world had to visit Barcelona. In Barcelona we considered the investment in Smart Cities the most direct way of regenerating the city's economy, in other words, to create employment. We were in the middle of the most difficult moment of the ongoing financial crisis, unemployment was widespread. What should we do? We decided to invest in urban and social regeneration as a fundamental method of achieving this. Today, despite all the regrets, which are more than a few, Barcelona stands out in the world of intelligence at the disposal of its citizens, and it leads the world in the rankings of creation of enterprises and initiatives whose aim is the development of content that betters the city and its services. A strategic investment, Smart Cities, generating work, attracting talent, improving the quality of life. That's our big secret.

MRS. CARME, OR SMART CITIES ARE NOTHING IF NOT A SOCIAL PROJECT

Smart Cities are cities that are prepared to adequately respond to the quotidian problems faced by the people that live in them, providing the best possible services, the greatest number of opportunities and the best urban solutions possible through the use of advanced technology. One of the solutions that has turned Barcelona into the world capital of cities since 2012 was the *Vincles*, or "bonds" project, launched that year by Barcelona Council within the framework of intelligent innovation initiatives and Smart Cities, and also by the emergency services (a great unknown that deserves a separate book). The project received the Bloomberg Mayors Challenge prize for the best intelligent cities project in the world in 2014. It wasn't based on savings, nor on sensors, nor on technological platforms. It was based around people.

Conceptually it was very simple, and therefore well thought-out, with profound implications: the project, aimed at elderly people who live alone, employs information and communication technologies in order to make sure that isolated elderly people, with few means, separated from the world, are never abandoned.

As I was saying, the principle was very simple, and it was the fruit of many conversations, investigations and considerations about how to make the most of big data, public and private cross-data analysis, the use of sensors, of remote care, and of the latest digital technologies in the world of health and intercommunication. The objective: to make the lives of the elderly people that suffer most, more dignified and safe. Nowadays we have the support of more than enough data and technology to ensure that the events of Reus do not repeat themselves. Creating algorithms that cross-reference the data of people living at risk, combining it with that which comes from sensors installed in the homes of those with a medical history, is at the fingertips of companies dedicated to the development of Smart Cities. Catalonia is the country that has the most start-ups related to big data. Who would have thought it only five years ago, when we set everything in motion?

Smart Cities allow us to find out if Mrs. Carme, 85 years old and living alone in a flat in the Plaça de Prim in Reus, or a flat in Eixample (the Catalan name for the Ensanche ["meaning widening"] district of the Old Town), is being well cared for. To find out if she is eating every day, if she opens and closes the fridge, if she goes for walks, if she speaks to her children or neighbours, if she takes her medicine, if someone visits her once a day; if, by extension, she still has a social and affective life. Smart Cities allow us to design the plazas where Carme can go out and soak up the sun, at the same time that a young mother plays in the sandpit with her daughter who's not yet one year old. Because one day they will be like María, active subjects of social progress. Reus was inevitable. We had to be more conscious of people and of development. This is the sign of the times, the authentic Smart revolution started in Barcelona.

In the world of cities, the innovative Smart concept launched from Barcelona was social but also corporate. It's the retired lady Carme, widowed and living alone in a flat in Ensanche; the manufacturing association in Ciutat Meridiana where young people from the city learn how to manipulate digital technology; it's the million pound investment by multinational companies from diverse sectors put into centres of excellence; it's the community of start-ups dedicated to the development of urban apps. The City Operating System is Smart, adequately evaluating public spaces, whether it's watering a garden or introducing the orthogonal network of electric buses. Here, there's little space for Manichaeism. Good and bad, black and white, half-told truths that adjust reality so it fits with the way you perceive it. Within this feigned contrast between the democratic and the intelligent there's a lot of it. In our country, unfortunately, we know what this entails.

If there's a human creation that is kaleidoscopic, complex, diverse, as some like to say, then that creation is the city. The city defines the world in which we live. We have divided the world into developed countries and developing countries. Nevertheless, what should we think about a city like Bangalore? One of the most technologically advanced cities on Earth and at the same time, one with abundant social needs. What about Shenzhen? Undeniably one of the most important economic hubs on the planet, one of the cities that has grown the most over the last few decades, while having inherent problems and an urban plan that we know will lead to social division. And what should we think about Paris? A city that was seemingly finished but that belies this idea every day in its *banlieues*. Perhaps the world is starting to divide itself into cities? Perhaps the world, more than into developed and developing countries, is being divided into urbanised and urbanising countries? The role Smart has to play in both environments is fundamental, as we are going to see.

Let's go to Mumbai, a city which which will be soon inhabited by more people than there are in the entirety of Australia; to Lagos, the largest city or human agglomeration on Earth, where hope is mixed with hell; to Los Angeles, to Bogota or to Barcelona, cities in which we can find the best and the worst of human nature, where we continuously come into contact with the limits of our technical and political capacity . . . wherever we go the city unveils itself as something real and as something with an intrinsic tendency for chaos. Smart Cities are the combination of tools that allow us to understand this chaos: managing, anticipating and adapting it (these are, after all, the distinctive features of human intelligence), assimilating its complexities and facing its challenges with objectivity. Smart Cities allow us to understand the city through high definition images. Using algorithms we design public space, we locate social services and we establish infrastructure.

Singapore, to give a weighty example, has equipped itself with a Smart programme on a national scale: Smart Nation. There, the idea of a city is confused with the idea of a state, it's true, which makes everything easier, let's admit it. The programme is based on a vision that aligns all universities, healthcare centres, schools, and training centres in general. It includes multi-million pound investments in innovation and development, the formation and consolidation of an extensive network of technology-based start-ups. The programme is protected by legislation favourable to the creation of businesses and through relationship channels with investment funds provided by the government, ready to be used to invest in the strategic focal point of the country. This focus guarantees the quality of life and the urban development of the city-state, with the objective of continuing to be one of the most important hubs on Earth. Furthermore, Singapore accompanies this idea with a well thought-out and aggressive system of internationalisation, of exportation of prestige . . . and of solutions. Creating, introducing, exporting. Another golden rule.

So, a Smart City is one that utilises resources (data, energy, processes, workforce . . .) in an intelligent way, it analyses them through information and communication technologies. In doing so it diagnoses problems, anticipates them in a detailed way and generates solutions with a technological foundations that can be scaled-up, reproduced and shared by citizens, businesses, and public administration. Because how would we be able we manage these cities with a population of one, two, three . . . or ten, twelve or fifteen million people, if not with the help of artificial intelligence applied to human intelligence? A Smart City has an intrinsic democratic base because for the first time, it allows citizens to be put at the centre of public policy.

Let's admit it, the Smart concept has been great for certain companies, because they've got an abundance of stories and catalogues for selling the unimaginable to many cities, that is to say to mayors, managers of services, council secretaries that see how daily life is slipping through their grasp. In this context, some have attributed all the evils of untamed capitalism to the Smart revolution. However, what we require, and what the companies that we all need, require, are mayors, secretaries, directors who are genuine clients, people that know what they want, but above all, what they want it for. Hence it is essential to situate public servants on the same level as their suppliers, so they become genuine clients, demanding and tough.

Only in this way will it be possible to improve services and make these providers truly contribute to the common good.

There is no turning back from Smart Cities. Without being the panacea, because at this moment in time we already know that there are no dialectics capable of explaining everything, the Smart Cities movement is the one that will allow working with and for cities, with the maximum possible potential for solving problems. The designation of a capable team and high-resolution diagnoses of urban complexity and intelligence applied to decision-making, constitute the only way for mayors to better serve people.

At some point we've all dreamt of being the mayor of our city and taking charge in order to "solve" the urban chaos in which we live. In the magnificent literary symphony that is *Maximum City: Bombay Lost and Found*, Suketu Mehta tells us that, tired of Bombay, tired of the populism of the left and the right, of tedious speeches, of the very good and the very bad, he dreams of waking up one day as the mayor of Bombay with absolute power. He dreams that he will be able to remove the Rent Act, a law that during the last sixty years has frozen the price of rent on buildings in the centre of the city, transforming it into a sack of battered and irrecoverable bones. He dreams that he will be able to give the judges back their judicial independence, without anyone abusing them or encumbering them with reality's overdramatic litigation, that he will be able to buy land to build thousands of blocks of public housing at an affordable price in collaboration with respectable developers and architects. That he will be able to develop small residential areas where children will be able to play safely, protected by a well-trained police force with authority and respect; children who are educated in high-quality public schools, with well-paid teachers, nearby healthcare centres, in a healthy and cultured environment. That he will be able to renovate the sewage and water management systems to eradicate infectious diseases . . . At this point, Suketu, and us with him, wakes up and thinks that all of this needs to happen in a democracy, yes, in representative democracy, within a legal framework, and that there is no other way to do so except by using all the technological tools we have at our disposal to make it possible. In the digital age the numeric, the interconnected, the technological are absolutely indispensable.

Woody Allen, who had an installation in one of Barcelona's cultural facilities rejected by the city, in a not very Smart decision, wrote a book called *Getting Even*, translated into Spanish as *Cómo acabar con la cultura* [*How To Do Away With Culture*]. Perhaps there's someone who would like to write a book called *How To Do Away With Smart Cities*. Those who think that we must immediately do away with Smart Cities can now be told that there's a way of achieving it. The concept will run out of steam once it is surpassed by another one that better utilises technology, that creates more opportunity, more work, better public spaces; that connects people easier, that ultimately is more resourceful and makes better cities and citizens. We predict that people will stop talking about Smart Cities when the management of a city has the Smart concept so assimilated and ingrained into its fabric that the "intelligent" requirement will be superfluous or redundant. The history of humanity is like this: only once we have mastered the technique of building arches and buttresses do we stop talking about the Romanesque, in order to move on to light and svelte heights of the Gothic. Only when Modernism becomes increasingly affected does Noucentisme (a Catalan cultural movement of the first third of the 20th century) begin, whilst the Avant-garde absorbs everything. We will be able to overcome the need to explore the Smart route, but before we do so, we need to find out all of its possibilities, exhaust them, and then start again.

In this book we make a call to action against prejudice. The city survives us. That human creation, perhaps the most human, the creation that speaks for most of what we are capable of as a species, survives us by explaining us. Through pieces of architecture we discern the languages spoken by the architects that drew and constructed them. In streets and squares we learn what the people that have given us almost all of what we have become, did for a living. In suburbs and waste grounds we read the residue of a civilisation that, despite everything, still has a lot of well-being to offer. Smart Cities give us these images, these readings, in a level of detail unimaginable until very recently. Moreover, they open the door to collaboration between all sectors of society, they shatter the monopoly on the creation of solutions.

From the perspective of our small and waning First World it is difficult to understand the extent to which the urban phenomenon is imposing itself everywhere, against all obvious logic. Cities grow and swell until they stop being cities and turn into camps of people that seem to be fleeing from something that we are unable to understand. We don't realise that the non-urban, away from the developed world, is deprived of the majority of the opportunities of the city. Nevertheless, we don't need to go to Ciudad Bolívar, in Bogotá, or to the shanty towns of Lagos or Delhi, to understand from what those who go to the countryside are fleeing. It would be enough to return to the shacks of Barcelona's Montjuïc, so well liked by locals and visitors. There, behind Can Valero, where couples get lost in the middle of nowhere to make love in secret, with the city at their feet, to recognise us amongst the misery of others and of the past. While there is no comparison with Mumbai slums, we, in Barcelona, two generations ago, had almost a third of the population living in slums in places that today are sexy beaches, sexy sports centers, and hidden parks for lovers.

To project and to transform are two verbs that mayors must know how to conjugate perfectly, to preserve the identity of the city and to permanently improve what it offers us. One of the best initiatives of cultural regeneration that we launched, with the invaluable collaboration of the City of Barcelona History Museum, was in the setting of the anti-aircraft batteries in Turó de la Rovira. That geodesic vertex, on which Barcelona defended itself from national and foreign fascist attacks during the Civil War, stopped being of military interest shortly afterwards, and was occupied by families displaced by destitution. A place, where previously shells had piled up, that was now piled up with people. A place where officers sought refuge, was now where children hid with their games, unaware of the way life pressured their parents. Or was it the other way round? Weren't those the parents that pushed themselves hard so that their children could have better lives? Today we have access to tools that couldn't have been dreamt of by people more than ten years ago. Why resist them? It's clear: be at the cutting-edge of innovation, and win. That's what Barcelona did, being at the cutting-edge of innovation and becoming the most innovative capital of the European Union. Accept innovation and survive, which is what is happening in a lot of cities and countries very close to all of us. Or, resist innovation and disappear. Innovation is also indispensable in the field of memory and identity recovery. Recovering the identity of the city limits, valuing the exceptions, in the places not reached by systemic urbanisation, is fundamental so that future generations don't forget the collective sense of urban life. That's what Smart Cities are for. If they weren't, we'd have to get rid of them.

There is a constant tension between transformation and being forgotten, and if we don't fight for memory, renovation loses its meaning. It's shameful to have forgotten the shacks of Somorrostro, the caves in Can Serra in l'Hospitalet de Llobregat, to have forgotten the Chinese district or the remote Singuerlín in Santa Coloma de Gramenet. A lot of digitalised adolescents have forgotten because previous generations fought to build a city without shacks, without foul-smelling streets and without epidemics and insalubrious spaces. In our case, Catalans from near and far. Some have idealised it. There's nothing worse than nostalgia for a landscape that never existed. Without doubt, in hindsight it's easy to be of the opinion that the Barcelona of the 1970s could have been better built. Digitalisation, virtual and immersive realities are tools that help us combat being forgotten. We're tired of them explaining everything to us the same way as they did one hundred years ago, when today it's possible to reconnect everyone to memory thanks to the digital world. If we're going to do it, let's do it talking straight.

AN URBAN PROJECT, TECHNOLOGY AND THE POWER OF TRANSFORMATION

To not recognise the transformational power of the urban project, and systematically oppose technological change for fear of the outcome, is the most reactionary thing that one can imagine. If everyone from southern parts of the world were to come to a Europe in which 75% of its population lives in urban space, it's because here there's an immediate answer to things that can't been answered in other parts of the world. But this is a story that must be put into the hands of our young people and their way of reading reality.

At the same time, remembering those shanty-towns that were so characteristic of the Barcelona I was born into, is a sign that not everything is lost, that it's possible to perceive change in so few generations. That the displaced people crowded together in Ciudad Equidad, in the Caribbean city of Santa Marta, Colombia, who are striving to turn their historic centre into the first Smart city centre of the Caribbean, perhaps will see how their hope becomes tangible in their city. We know that the speed of change is accelerating, and we know that our capacity to react diminishes if we don't use tools that work at the speed of light. If we do it well, thanks to education, the digitalisation of reality, and the integrated management of the city, we will see how, in a short time, our capacity to deal with challenges that have been impossible up till now, will improve. Challenges like piles of rubbish, the standard of drinking water, the distribution of energy, or the management of traffic. This is the position of cities like Pune or Nagpur, in the Indian state of Maharashtra, for example. Cities that don't want to be marginalised, cities that aspire to produce a complete life experience, and that use their Smart strategy as the way of doing so, to drive the reforms that will permit equality of opportunity to all of their habitants.

The acceleration of the urbanisation machine creates other challenges. In Pune, Cairo, or in Lagos, refuse is processed and recycled by hand by brigades of displaced families or Untouchables (which is the same thing) that live among the waste of others; their life expectancy doesn't reach forty. Whilst, incredibly close-by, in the most expensive condominiums on the planet, or in state-of-the-art universities of technology, from which the best IT and telecoms engineers graduate, tomorrow's millionaires are being shaped. Inequality is not a new phenomenon, but the distance between the extremes is growing and mayors are not born educated. Many are powerless before the phenomenon.

In Lagos, the biggest metropolis in the world, a six-mile journey can take more than three hours. Similar things occur in Rio de Janeiro and Beijing. In these cities helicopter businesses prosper; they give a symptomatic solution to the permanent grid-lock in which half of humanity is immersed. For some people this is the most intelligent solution. Others think that taking the cars off the streets will solve the problem. But we don't believe in one solution or the other. Urban intelligence exists when it is social, that is to say, when it benefits everyone equally, taking into account the needs of those who have the least. Urban intelligence demands cross-referencing data, analysing behaviour and, whilst structural solutions are being planned, solving problems whenever they reveal themselves.

It's fair to say that, in light of the problems that have arisen, some people advocate neither "smart," nor, "cities", what they think we need to do is to scale down, to make ourselves smaller . . . But the truth is, in the real world, cities are often the best alternatives to a bad bunch. People flee the misery and violence of the Tanzanian and Nigerian countryside, and cluster in endless Lagos or at the gates of Dar es Salaam, looking for a vital alternative that, if it doesn't meet expectations after all, will at least save them from fatal suffering. Cities have the potential to be the solution, even if we don't rise to the occasion. In Tanzania death continues to stalk its youngest: the entire country has barely a handful of paediatricians at its disposal. In developing countries people go to cities as a first way out, but soon they realise that these new megacities can be as cruel as they are discriminatory. They then become merely a stage on the road that will take them to any "developed" city. Let's not say in bombed Syria or in missing Libya. Let's be clear, great migratory movement has only just begun, and it will affect in a decisive way what we are going to be in the future. It will depend on whether we realise the extent to which the phenomenon is irreversible, so that we prepare ourselves positively for it. Preparing ourselves means preparing our cities to integrate them into a very different world from the one they come from. Cities can be hope, those who arrive might bring the knowledge we need to rejuvenate our ancient continent, but to do so we need mayors prepared to transform and capable of creating opportunities for everyone. No-one is born taught and everything is still to do. So, in cities that are growing at a rate of fifty, one hundred people a day, how are we going to deal with the problem? How can we plan without wasting time? Could the solution be Smart? Could the solution lie in the development of technological solutions, based on a shared will?

Smart Cities might be the salvation of those cities that almost were, but that are not because no young people want to stay in them. In Catalonia we have examples of this. Take the Urgell axis, which descends from the regions of Pallars to la Segarra: Tremp, Ponts, Artesa de Segre, Balaguer, Tàrrega, Cervera . . . Cities that have been cities, even with universities, that today struggle to preserve their historic quarters, to recover their battered heritage. Tourism passes them by, because they're too far from the sea, or the snow . . . and Barcelona is too close. Smart Cities should give way to Smart Villages, because what these small towns still possess is a tranquility and relationship with the natural environment that is the envy of larger cities. Likewise, all of them still retain some commercial and even industrial activity, from which they can develop a Smart positioning strategy, without intermediaries. There is a regeneration proposal waiting for each of them, and for them as a whole. What we need to do is approach problems with a 21st century mentality, fleeing from territorialisation, industrial estates and destination.

Technology has been a driving force for change since anyone can remember. In the 19th century, the possibility of channeling water (it's essential to visit the Urgell Canal Museum, in Mollerussa, to understand this), deploying sanitation networks and building with steel and glass exponentially multiplied the ability to make cities. It was the age of ambition for "cities of tomorrow": expansion, garden cities and tall buildings. The 20th century revolutionised mobility, invented sustainability, but only half beat the temptation to create a suburban world. Today, in a world that is bursting at the seams with unemployment, lack of opportunity, mass displacement, violence and inequality, it's obvious how technology works to help people regain hope.

The internet has changed and is changing the world. Not only from the point of view of how we communicate, but from how we live. Digital transformation is developing before our very eyes, without us having to do much to take advantage of it. In the year 2020 one single family will generate the same volume of digitalised information as the entire Earth in the year 2008. That power cannot be unaffiliated with the city phenomenon. At the end of the day, citizens themselves have brought their smartphone to every corner of the city and they're the ones who have integrated connectivity into their way of life.

The current technological revolution has been far more democratic (yes, democratic) than all of the revolutions we know about from previous centuries, some of them only a hundred years ago. The introduction of digital technology into daily routines has not been a planned imposition from a five-year plan, nor from an engineer's office, nor has it required armies of operators with helmets, or concrete plants. The change in paradigm has arrived silently, without any more infrastructure than antennas, the fibre-optic network, and the terminals we all carry. Human beings have understood that digital tools allow us to radically improve our quality of life. If you don't believe so, ask the millions of Africans who today make payments through telephones, or the students of the digital universities that allow them to learn in any corner of the world, if someone has had the decency to install a WiFi connection. We live continuously with a window to the hyper-connected world in our pockets.

This is why Smart Cities are a provocation, because those who haven't wanted to see them have been able to deny their existence . . . until it became too obvious: Smart, digital, is here to stay, because it's a tool that can objectively improve our lives and provide answers for real and pressing human challenges. However: being Smart is not simply accumulating data or information. What's Smart is dealing with that information in favour of the construction of a healthy, responsible citizenship, implicated in and with the community. What's Smart is growing without voracity, becoming aware of the value of interdependence between people and their social and natural environment. What's Smart is closely linking the social, the economic, the cultural, and the ecological. What's Smart is using these tools of mass relation and creation, digital tools to benefit human beings.

Smart Cities are not, therefore, a technocratic decision, but the consequence of radical change, with no going back, in individual routines oriented towards common good. It's citizens themselves who have become accustomed to reading at all hours of the day, interacting through social networks, buying with a click, coordinating teams through email. And they don't understand why mayors don't use the technology that they use, with a single finger, to better serve the city. The potential is so obvious that it's impossible not to transfer digital technology to the transformation of the city. We could live without Smart Cities, of course, but not for long. And we'd be worse off.

Good mayors are generous with future generations, they think beyond their mandate and listen to the beat of urban life. Smart Cities, as we understand them, are oriented towards preventing the impending doom that dominates the worldwide urban phenomenon. The utilities revolution is about to arrive and will offer new and better services for the management of energy resources. The digital revolution of urban space is taking place in cities like Barcelona, Copenhagen, New York, London, Doha, Pune and a few others. But today the demand for producing top-quality shared space, sensorised and hyper-connected space, is already commonplace. Now, do we know what we have to ask the digital world for in order to make us and our cities better? That's what I propose explaining in the pages that follow.

Cities are obsolescent factories of living. That's their main problem and their main opportunity. If our great-grandparents were able to imagine and create a new city, in the broadest sense of the term, based on technology available at that time, why shouldn't we be capable of doing it? Protagonists as we are of the most significant scientific and technical revolution in the history of humanity. As my old friend and comrade in arms Dr Manel Sanromà always says, the commitment to

Smart Cities is precisely the answer to this revolution. As with the start of every intellectual revolution, the first steps stutter, there's a lot of noise and everything's confused, social urgencies seem to have no respite, industrial supply is diffuse, the inability to ask the model what's really important slows down evolution . . . What's necessary is reflection without stopping the machinery, because the city will not wait.

We contemplated all this in Barcelona a few years ago, when a group of people realised that we needed to revive the city to create quality employment, to retain and attract talent, to raise investments. The Olympic spirit had vanished and the crisis was hitting hard. The industrial world, the world of knowledge and services, resisted, but it was our responsibility to make good use of the city and turn it into the most useful tool to escape from that economic slump. From here on out, everything else happened. I mean to say that the Smart revolution in Barcelona happened simply because we proposed it, without further ado. Because, ultimately, it's a revolution that can take place anywhere on Earth, as is happening, because it only depends on three factors: having a city ready to accept the challenge, having a vision of improvements that can be transformed into a plan of action, and having a team at your disposal that is prepared and eager to lead and control the transformation. I will talk about all of this in the chapters that follow. I will do so by filling them with history and stories, with that component that makes some cities more magical or sexier than others: strength of identity, something that in Barcelona we are not exactly lacking.

3

Barcelona: A Mantra, A Plan

TOWARDS THE NEW BARCELONA MODEL

Some time ago Barcelona developed the urban model that made it famous in advanced circles of architecture and international city planning. The model was designed by the Barcelona School of Architecture, lead by Oriol Bohigas and set in motion by the first democratic local governments, using Pasqual Maragall (previous Mayor of Barcelona, who won the 1992 Olympic bid) as a point of reference. The model involved promoting the values of the dense and compact city, its Mediterranean mould focused on cultural heritage, contributing social values and regenerating the rich identity of the impoverished urban centres caused by the irresponsibility of some, and the rushed planning of others. The urban acupuncture of the start of the eighties, the building of monuments and the improvement of the suburbs, combined with Olympic planning, turned Barcelona into what it is today: a fundamental benchmark for global cities in matters of restoration of public space. We are the children of this story.

Over the years, once the sweet taste of post-Olympic successes had diminished due to the manifold crises that the city experienced, some of us proposed reforming our ideas, visions and plans of action. The city couldn't carry on living off a vision developed during Franco's regime, many years before any hint of the digital revolution. In fact, what was needed was an update of the model, based on more deeply structural elements than the ideas that were contemplated in the eighties. The worldwide economic crisis is rooted in the demise of production, energy and financial models, and Barcelona was not immune to it. After the property bubble and overwhelming level of unemployment in the city, we decided to broaden our horizons. We needed a new vision that would guide us over the coming years. A mantra through which we could evolve, without losing the essence of social commitment, of the social dream that Barcelona has been for more than a century and a half.

We decided that Barcelona should become a global leader in the restoration of public space, but also of domestic space, its productive capacity and energy self-sufficiency. I led an enthusiastic and intelligent team that set upon the task with conviction. Barcelona needed to emerge from the crisis by brazenly committing itself to the new economic revolution dominated by artificial intelligence, and applying it to the city, regenerating it, which in turn amounts to regenerating society, culture and the economy. We got together to offer solutions to the systemic problem that is the obsolescent city, unable to offer housing and decent work for all, and we used the city itself as that solution. Ultimately, Barcelona has managed to overcome social problems with dignity thanks to an equitable city model, based on the things we share, at the same time as promoting civic and commercial initiatives.

Another of the key elements involved in taking this decision was observing what had happened to the "Barcelona model." Studied, known, highly regarded . . . but little else. The same thing wasn't happening with urban development models in Singapore, Japan, the United Kingdom or the United States. These countries, equipped with well-constructed and well-deployed strategies for international expansion, thanks to efficient diplomatic services and a strong export-focused mentality, were imposing themselves all over the world. Barcelona, on the other hand, was solely present in academic circles. By travelling a little, we would discover that Barcelona's model definitely hasn't prevailed in the corners of the world where new cities are growing, principally in Asia, South America and Africa. We would discover that the cities that are being built today, wherever they may be, follow parameters that go radically against those that we believe are the defining features of our successful model. Servitude to cars, a lack of attention to energy efficiency, Mannerist architecture, contempt for identity, the zoning of land, a lack of respect for precedents . . . These competitors are imposing themselves on us, they're beating us on all fronts. Who are they? They're the urban planners, engineers, bankers, factory workers and architects from all over the world that, accompanied by a system of active diplomacy which allows them to develop their models from Guangzhou to Astana, and Dubai to Johannesburg, continue to impose their utilitarian frameworks. Frameworks which prioritise urgency and short-term functionality ahead of flexible, long-lasting growth that is adapted to local conditions. Frameworks that condemn these societies to profound identity and socio-economic crises. All of this distances them from our conception of a city. We decided therefore, to combine the mantra with a large-scale strategy of globalisation, of giving a practical importance to our urbanism and capability of planning cities. This could only be done through a commitment to social and transformative Smart Cities, based on an urban project in the style of Barcelona. I equipped myself with a team, a mantra and a plan. It was as simple as that.

CORTÉS, MICHELANGELO, JULIUS II, CERDÀ AND SMART CITIES: RESPONSIBLE SUPPLIERS, INTELLIGENT CLIENTS

Nevertheless, to arrive at this mantra, plan and team we needed long-term vision and decisiveness. Years before this, together with Vicente Guallart, Willy Müller and Manzuel Ganzo, we founded the *Institut d'Arquitectura Avançada de Catalunya* [Institute for Advanced Architecture of Catalonia]. In other words, I surrounded myself with people that were prepared to break the mould, to burn bridges, to risk undertaking an ambitious intellectual venture with practical consequences. We collated the best things about our tradition and we threw away what we thought wouldn't be useful to us. We opened our eyes to the world. It was entirely a conscious decision, thought-out, argued and planned. I read everything I could, and more. We found our inspiration in stories from history, and I stumbled upon some characters that have accompanied me for many years: What do Hernán Cortés, the Spanish explorer; Ildefons Cerdà, the father of urban planning as we have known it for the last fifty years; Pope Julius II and Michelangelo have in common with the world of Smart Cities? Well, a lot more than it seems. The unknown creates spectres, and all that glitters is not gold. We soon found out that what happens to many cities with their star architects, large businesses, and suppliers of management technology systems is similar to what happened to Hernán Cortés: a few horses, arquebuses, tawny beards, small mirrors and bejewelled bracelets were able to topple an empire. Cortés had clear ideas: when he arrived at Veracruz, after drawing his famous line in the beach's sand, he scuttled his ships. His objective was clear: plunder Montezuma's gold at whatever cost. His destiny was coloured as the setting sun, and towards it he marched. Little did the intrepid Extremaduran know that Montezuma's great empire, sick with superficial bureaucracy, social stagnation, internal division, cruelty and technological underdevelopment, would so easily succumb. Cortés and his men crossed the many territories of the Aztec Empire until arriving at Tenochtitlan, a city which, viewed from its hillside surroundings, had a central square larger than the one in Valladolid, according to the exquisite description by Bernal Díaz del Castillo. It was already all about cities, and Cortés, a man of the court, knew that his game was played out on urban terrain.

Hernán Cortés wasn't the only person to confuse his adversaries with jewels. British and French explorers advanced through the valleys of the Yukon and the Rocky Mountains trading hides for stories, territories for alcohol. The North American aborigines were not aware of the destructive potential of this exchange . . . they were equally unaware of their own potential, which combined with that of the territory's new tenants, perhaps could have improved their well-being to unimaginable new heights. The problem? The avarice of the conquerors and explorers, and the aborigines' ignorance. Sometimes a similar thing happens to suppliers and cities.

Some of the catalogues of elite architects and companies that supply Smart solutions for cities contain many of these little magic mirrors: They promise effects similar to the ones produced by Harry Potter's wand and they forget how important it is to understand their *client*. To put themselves in the client's place, to empathise with them and to do it all sincerely, given that what we're talking about is real interest in new solutions. When the Aztecs realised that Cortés and his men were not gods but rather human devils, as human as themselves, it was too late. The conquistadors had sold them objects and a story that was useless to them: to them and to their gods. By the same token, the assortment of sensors and screens promised by some might not be anything more than a mass of iron and plastic installed in our streets, if the mayor and his teams don't know exactly what they want. Thus, the fundamental challenge for all those who are responsible for a city should be to become excellent clients. People that know what they want because they have thought about it, because they are looking beyond the present day. People that have made a correct diagnosis. It's easy to say, but very difficult to achieve.

I would like to make it clear that we have a lot of respect for people who believe in the fruit of their work and muster up the courage to promote it. We don't share the view that demonises businesses and businessmen; we believe that the private sector and well-regulated market can contribute to the development of a city, if from within it citizens' priorities are properly communicated. Supply, demand, and solutions, yes. Because the story is about clients. The problem is that we don't know how to be clients a lot of the time, because we haven't learnt to know what we really want, what we can get out of companies that, with the ability to help us, exclusively try to make as much money as possible. Linked to the ignorance of public decision-makers, mayors, deputy mayors and senior officials, is the inability to be good clients. Every supplier that values itself needs, in order to improve, a good client. This is the case with Michelangelo Buonarroti and Pope Julius della Rovere: two strong personalities, two men convinced that they were in permanent contact with God . . . and with his immense humanity; two figures that did not exactly like each other, but knew that they needed each other. Neither one of them would have had the Sistine Chapel without the other, the work of art would not have been painted without the commission (or the political, economic and social pressure). Client and supplier, therefore feed off each other when one knows what they want and the other knows how to achieve it, how to get it off the ground and how to listen (even if it is through clenched teeth).

Something similar happens with Smart Cities: we're tired of seeing pilot projects that consist of nothing more than quick-witted sales teams. However, we're also used to seeing situation rooms that are neither rooms nor do they show any signs of life when a situation arises, simply because nobody pays them the slightest attention. Often I remember how, during an official visit to South America, a mayor who was very proud of one of these rooms, of *his* room, emphatically insisted that we should visit it. We left the magnificent city hall which was French in style, with large ballrooms, conference rooms, dining rooms and representative offices, amongst them the mayor's, and we headed to a new building that was about twenty minutes away by car. Here we met with the manager of the city's traffic control and emergency response centre, a man accustomed to these types of visits, and he started to explain a bit about his work. The presentation changed tone when we were directed to the situation room. The manager left himself at the disposition of the mayor who asked him to put the entire potential of the room on display. It was obvious that it hadn't been used in months: dust on the tables, disconnected screens, ageing computers. The most important thing, however, was yet to come: the mayor's screen. A screen that was connected to his office in the romantic French-style building, that allowed him to follow non-existent meetings, as the manager himself confessed to us later on.

The danger of large companies on the hunt for an unsuspecting mayor is that they might fill the streets with sensors that don't do anything, and offices with screens that do even less. We've experienced it in our own city: when we proposed installing an accessible integrated system of sensors for the entire city, the forces of indolence, ever present in our large organisations, tried to convince us that it wasn't necessary to advance so far, as we were already sufficiently equipped. In fact, they took us to see one of the latest refuse lorries that we had acquired. It had a hybrid combustion and electric engine, kitted out with all kinds of signal detectors that in theory would receive signals from the sensors installed en masse in rubbish bins spread across the city. Everything was prepared to triumph that day. What I mean to say is that everything was prepared so that the "we already did that" or the "we already tried that and it didn't work" bureaucrats would triumph. We got on to the giant toy that a cutting-edge rubbish lorry feels like, and they sat us next to the driver, to whom I'll never be thankful enough. His name is Diego, and he was a source of inspiration in demanding that inventions have real utility. The demo consisted of situating ourselves next to a full rubbish bin, detecting it with the lorry thanks to the sensor that indicated the state of the bin, which in turn automatically activated the collection system. I was alone with Diego in the cabin, and soon we started to bond. I told him that one of my grandparents had been a street sweeper for the council in the middle of the 20th century, and that we'd always felt proud of it. Bringing out the best in Barcelona is one of the most noble occupations that one can have. Even more so now that we've moved on from the simple broom to the most state-of-the-art technology. Whether it was due to the feeling of proximity, or perhaps more likely to Diego's professionalism, we proved that not everything was as Smart as we had been led to believe by that display of resources. "Do you want to know which one is my sensor?", asked Diego with a smirk, after I had sung all the possible panegyrics about the sensorisation of the city. "Look, Mr. Vives, I get close to the bin, in which of course the sensor doesn't work, or exist, whatever they tell you. I lift it up like this, see?" Diego activated the lorry's system, which clipped onto the bin and lifted it a metre off the floor. "Now I shake it like this", he smiled. "If I feel like there's rubbish inside I empty it. If there's nothing in there, then I leave it for next time. There aren't any sensors here, the route is what's important and that's it". You can imagine how foolish I felt. You can also imagine the immediate analysis that we started from that day on, analysis that made easier the implementation of the platform that integrates the open data from civic sensors, the *Sentilo* cross-platform. Until that moment the existing system of sensors had been conceptualised for initial pilot schemes, but not for taking a decisive step towards improving the quality of service, or the quality of Diego's job. What was important was the pre-established route, the number of rubbish lorries on the streets, the number of street sweepers . . . We started to work on changing the paradigm, using the power that Smart gave us.

Clients, good clients, responsible suppliers, less worried about cashing in a cheque as soon as possible and more about how to provide a good public service. Fewer little mirrors and more compromise. Fewer dreams of grandeur and more awareness of what we want and, above all, why we want it. This takes us back to Cerdà, whose story everyone, or almost everyone, is aware of: Cerdà is the king of the self-commission, the inventor of a client that up until that moment did not exist: the city. I'm talking about the meaningful self-commission, because the trick doesn't always work, as I will discuss later on when I talk about New York. Ildefons Cerdà (a 19th century urban planner) succeeded in prevailing over his competitors because he knew the secrets of the Madrid administration. Barcelona's Ensanche district would have been different had it been proposed by the city itself. It was submitted by the architect Antoni Rovira i Trias (with the inspired and enduring motto *Le tracé d'une ville est oeuvre de temps plutôt que d'architect* [the outline of a city is the work of time, not an architect]) to the tender announced by the council on the 15th of April 1859. On the 19th of November of the same year he won. He proposed a radial structure formed by six grand avenues that, by building on the existing city centre, would connect to the suburban districts (Sants, Sarrià, Gràcia, Sant Andreu and Sant Martí). The spaces in between these large avenues would be subdivided into blocks with different sized edifices, at the same time as looking to group them into neighbourhoods through the strategic placing of plazas. To the north of the Rambla, a grand plaza would be the element of union and interaction between the two urban fabrics. The projected expansion would be surrounded by a new train line that connected the three existing ones, and concurrently a canal would be planned following same route to divert the rainwater that flowed down from the mountains in Collserola to the sea. In the pre-existing urban scheme the construction of a new street that would lead to the cathedral was proposed.

The common denominator shared by the major plans that were being proposed in Europe at this time (the Vienna Ring Road [1857] is one of the most representative) was to find the maximum possible connectivity between different parts of the city, to establish a certain hierarchy between districts and to contemplate the city as a nest-like object with limitations. Rovira i Trias solved Barcelona's specific problems by following these ideas, through the radial layout of the main roads, the placement of official buildings in important places and the establishment of new clear and defined city limits. For all these reasons it was the most coherent of the projects that were presented, it tried to adhere as best it could to the guidelines set out by the council and to prevailing theoretical principles. However, according to the Royal Decree on the 31st of May 1860, the project that would mark the future of the city's growth would be the engineer Ildefons Cerdà's, the fruit of theoretical principles based on a limitless vision of the city. It was a plan without hierarchy that contained homogeneous distribution of activity.

I don't believe in the goodwill of a few seasoned Spanish civil servants, who are always prepared to stop Barcelona completely having its way. Rovira i Trias' project contained aspects of the era's pretentiousness, comprehensible for those who were used to the precarious theory of urbanism that was valid at the time . . . This must have alarmed them. What? Barcelona memorialised? Barcelona filled with major avenues and space for large sculptures and royal celebrations? The only way of stopping it was finding an alternative that could replace it, also developed by a Catalan, but without the delusions of grandeur of the city's project proposal. Cerdà's project fit like a glove: egalitarian, isomorphic, although some considered it depersonalised and overly scientific. In the words of Josep Pla (a contemporary Spanish artist), an urban encampment truly lacking any personality. This was the winning project. This led to the Spanish authorities infuriating Barcelona for the thousandth time. This was one of the reasons why Barcelona is recognised today as one of the most intelligent cities in the world. Cerdà invented a multilateral, neuronal and fractal hive, something like Ramon Llull's *Tree of Science*, another madman of our homeland, transferred to the urban world.

Barcelona was furious and started to make the the most of Cerdà's proposal. Cerdà's shrewdness was that he considered the technique and technology of the time his ally. The symptoms started to become clear: an improvement in hygiene, the symbol of which was Monlau's project *Abajo las murallas* [Down with the walls] in 1841, the introduction of new technologies, whose effects started to be felt in 1848 with the arrival of the Barcelona–Mataró train line and the creation of the Industrial Institute of Catalonia. Mass transport and industry in the city: two driving forces that Cerdà highlighted in his project, and that we have recuperated in the 21st century. Don't forget that Cerdà was a military engineer, and despite the fact that the Spanish military had not won any battles outside Spain for a long time, it didn't mean that Cerdà wasn't in the know. In effect, Cerdà was a Smart engineer *avant la lettre*, and under his guidance Barcelona became a Smart city in due course, defining an era and a way of doing things whose echo has even reached our time.

Cerdà's Eixample and his theories on the city are situated in the context of the political, professional, social and economic transformation of Catalonia and Europe during the 19th century. Cerdà was a by the book modernist. This was the period of the revolution of ideas, the age of the great critical thinkers, from Marx to Nietzsche, including the first serious attempts at modern reform in the heart of the Catholic church. This was the time when the political principles borne out of the French Revolution and the romantic revolutions of 1848 were put into practice. The dawn of national identities that had been repressed throughout history. They were the years of the blossoming of utopian socialism, and Cerdà, a man of the times, well equipped with ideology, ideas and technical training, invented the client. No cheap trinkets, trying things on a small scale or taking commissions just to scrape by . . . Cerdà invented himself a super-client, and he got to the job.

It's easy to agree on the priorities we want carved into the pediment above the door of our city hall. And yet, as clear as it may seem, it's a nightmare to bring about, and what's worse, to explain and talk about. The truth is that everything is made up. When I started to work in a factory that made cars as a young man, it really caught my attention that the majority of employees, and without doubt all of the managers, wore a tag that hung from one the pockets of their work jacket. It read "TC3". After a few days I was wearing one of these tags as well: "TC3: Taking Charge of Customer Care." This tag reminded us to share the responsibility of taking care of the intermediate and final clients, external clients, but also internal ones. The company's quality started on the product design table, followed by management, the purchasing department, engineering, finance, marketing, sales, after-sales, training etc. We were all TC3, even the people that usually made fun of these types of campaigns with a smirk that's so characteristic of our neck of the woods, that doesn't show approval or disapproval, in fact quite the opposite. Well, even these people, in the end, knew what to expect. Taking after the CEO Joan Amorós' apt expression, in our company we applied the *progressive sales* concept. (Amoros is president of several Catalan cultural and economic associations.) A concept that started with letting the client know exactly what the design criteria would be, including production, logistics, sales and post-sales. Could we do the same with citizens and the city?

THE MANTRA

We've always been convinced that it's not good enough that only a few people have access to the core of ideas and principles. In an administration, corporate principles also exist for the benefit of the people. Therefore the first thing that I did when I started in Barcelona's council was to make sure that everyone that worked for me knew what principles to abide by. We invented a mantra from which everyone had to creatively develop abilities, a TC3 for Barcelona borne out of in depth observation and permanent dialogue with the city. The mantra was as follows: **Barcelona will be a city made up of districts that produce at human speed, that are entirely energy self-sufficient, hyper-connected and produce zero emissions.** We received all kinds of comments from the opposition, the majority were condescending. But now, when people ask themselves why Barcelona became the global reference point for

Smart Cities in such a short period of time, why Barcelona was able to become one of the first world cities to attract talent, creating jobs associated with reforming the city. Why Barcelona won prizes, capital and investments, why Barcelona reformed the Avinguda Diagonal, demolished the Glòries ring road and created the network of manufacturing associations, everyone was clear: synthesis works. We were, and still are, convinced that this mantra was the fuse that sparked the new leap forward for our city. And we are convinced that, without politicians, social leaders, businessmen or citizens prepared to work based on ideas, amassing them, dedicating them time to bring them to fruition, constructing them from the inside out, and not the other way round (we're tired of dealing with people that only repeat things, without ever letting anyone know what they really think), it's impossible for society to advance.

Barcelona lived immersed in the unacknowledged fatigue of the millionth by-product of the Olympics. The Forum of Cultures extended the Olympic dream's zone of comfort, a dream in which the majority still lived. Just a minute: Barcelona had always been able to take steps forward, and at that time, although exhausted, it still gave us the incinerator and the cogeneration plant at the mouth of the river Llobregat. The stimuli, however, had run out. Barcelona needed that new impetus that moves it along every so often. Before, it was every two hundred years, perhaps more. Then it was every hundred, every fifty. Today, at the apex of Barcelona 5.0, it's every thirty years or less. It needed something fresh, bold, ground-breaking proposals, and we put ourselves to it.

The objective, to create a new narrative, to attract investment by recuperating the industrial soul of our city, and with it, the soul of all the cities of the world. To create high-quality employment. I must emphasise: high-quality employment. It was a long-term commitment. A city is production, creation, industry, as well as culture and identity, always with the goal of providing everyone with the most comfort possible, without forgetting anyone. We needed to be able to provide something fresh, something new that was based on our profoundest convictions and our vast experience (that quality that is so often despised by those who rave about new things with no substance).

In order to do this we needed to tackle basic aspects of our way of creating society. In essence, the energy model. Cogeneration had paved the way, but it wasn't enough. From this idea sprung our obsession for energy self-sufficiency. Today Barcelona is on the cusp of achieving it. The only thing needed is determination and clear ideas. Self-sufficiency was seen as impossible, even by the most progressive members of the council. When I would ask them why, they'd respond, "Because it's impossible", without going further, without realising that this fuelled our desire to achieve self-sufficiency even more. We'll talk about energy, but it needs to be made clear that our commitment forms part of a way of seeing the world, that we don't agree with the way we import well-being in the form of petrol, and that we see ourselves forced to export instability. It can't and must not be like this.

All of this is included in the mantra. The mantra has to be the product of individual and collective reflection over a period of a few months, if not years. It took us ten years. Because the city is an artefact of periods of long-term development in which, sometimes, the effects of investments are not seen until some years later. Without a mantra, cities, like any organisation, never establish their karma. Another matter entirely is that the city has to feel like change cannot be vouched for in the long term: a long term-outlook is necessary, but also a good plan of early victories within the first six months. Five-year programmes combined with bi-annual programmes.

THE STRATEGY

We've already talked enough about the mantra. Because a mantra without a deployment strategy is entirely useless. My strategy was focussed on three key areas:

1. Organisation. We created the Urban Habitat unit in which we merged: housing, infrastructure, parks and gardens, waste disposal and management, urbanism, architecture, metropolitan urban planning, electric mobility, water, electricity, digital and communication technologies and the international promotion of all of the aforementioned. There wasn't a single urban project that didn't take into account an integrated approach to all of these aspects. My general staff developed the urban transformation from a holistic perspective, integrating it at street level into the urban project.

2. A long-term plan with a good combination of objective indicators, associated with the big political goals and improving people's quality of life. However a project should also incorporate a scheme of early victories, within the first six months. In other words, the vision, the mantra could only be seriously considered if we were able to change aspects of the city within a short timeframe, with high-impact and low-cost operations. Hence, micro-urbanisations, the apps strategy for Barcelona, the implementation of vertically integrated Smart solutions like public lighting or risk, etc. were converted into undeniable factors for widespread recognition.
3. Global branding and positioning. We were convinced, and we still are, that we had things to say to the world that would positively transform it, but above all, that would support the capacity of our professionals and businesses in a time of crisis. To be able to say things to the world you need to form part of it: this is the reason for the bold investment in internationalisation, the Smart City World Congress, the Mobile World Congress, in the conferences and international agreements: we attracted the World Bank, we allied ourselves with Hong Kong, we attracted GWOPA (Global Water Operators' Partnership Alliance), and the United Nations and its urban resistance unit. We conceptualised and created the City Protocol, an open source code for the development of public services in all world cities. And we created the City Protocol Society, to which more than forty cities from all over the world, universities, and businesses interested in popularising the Smart concept signed up. The anatomy of a city scheme, developed for the City.

Therefore we knew what to do from the start. Vision and plan, that was our secret. What to do? That's the big question asked by so many mayors, public decision-makers facing a task that doesn't seem to have a solution. That's the question asked by so many citizens like us that helplessly assisted the triumph of the error. It's also the question asked by so many consultants, by heads of small and large businesses who, without having had any type of public responsibility, hope for answers without having formed the right questions due to lack of knowledge or the inability to understand who the real client is and what they expect. To paraphrase Camarón,[2] (footnote overleaf) life in the city is

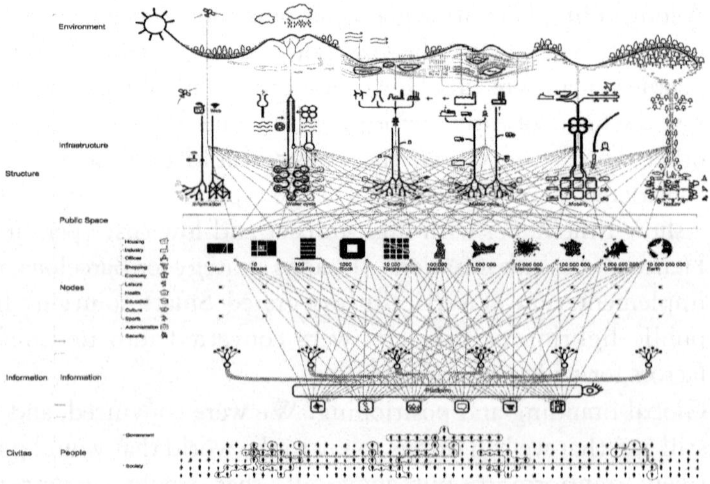

City Protocol Society idea, by Manel Sanromà and Vicente Guallart, has already become a classic of civic open source. In short, by exposing ourselves to the world we attracted, shored up and exported investments and talent. And talent is the most valued resource in any society.

an obstacle, it's in constant flux: an amalgam of the physical and the human, of the empirical and the spiritual. Digital technology adds a more complex layer to the city, making it even more interesting. Our project for Barcelona was a project open to change, and the commitment to Social Smart Cities is framed by the choice to uncover digital technology's maximum potential in all areas of urban life: from social relationships and the economy to culture, services and the physical transformation of the city.

The value of change and the step up in scale of Smart Cities in Barcelona resides in the multidimensional nature of the programme. The moment businesses started to develop technological solutions to improve urban services, we committed to declaring Barcelona a "city open to change" and guided the corporate ecosystem (from local entrepreneurs to multinationals, from schools to universities) towards solutions that were truly useful for residents. The ambition, the yearning which inspired us, was to convert Barcelona into a city that produced solutions, instead of simply a consumer of them. Thereby the city would become inherently a new economic engine. In the end,

---

2 Considered one of the greatest singers in Flamenco music, 1950–1992.

cities are obsolescent tools for cohabitation and they need to be constantly brought up-to-date.

We decided to convince the world that Barcelona was *the* city par excellence: historic and modern; elitist and popular; conservative and avant-garde; clerical and profoundly secular and transgressive; productive, a tourist destination and a service economy . . . The indisputable capital of Catalan sentiment, and precisely for this reason, cosmopolitan like so few other world cities. To a great extent we achieved it. Today, more than at any other time in history, we talk about and work on cities: the regenerated, redeveloped city and the investment in innovation and digital technology that generates a bank of new plans and projects that have been able to turn cities around. This new type of city that we launched to the world from Barcelona is able to share learning curves for innovative projects with other evolving cities.

However, it must do the same thing as any other world city that doesn't want to be overcome by stagnation, or by the last distant disciple of Albert Speer who strolls through the arcades of the architectural franchise markets. Stagnation also hides itself behind Neo-Luddism, a movement that bases its opinions on apparently political precepts, but in reality embodies long-established ignorance exported to the world of prejudice. The only way of overcoming stagnation is with tangible projects that truly transform, successful cases that incite change and transformation in ways of life. Barcelona's challenge, one that we pose for all world cities, is developing a city with vast public space, recovering the initiative to create housing for all, equipped with a productive industrial ecosystem that forms part of its urban nature. An ecosystem that is implicated in the development of a self-sufficient city, with zero emissions and technological intelligence. This city must be a reality within two generations, but it must start to become it from right now. In fact my proposal is to do with the civic combination of the structural and the immediate, that which has to survive us and that which will make our lives easier tomorrow.

Therefore Barcelona's challenge was, and is, the challenge of the Western city, and in essence the challenge of any city of the world: to stop declining, to stop languishing in order to be able to renovate our streets and squares as cradles for the creation of opportunities and social and economic wealth. The architecture, design, urbanism and engineering that must support this model have to be state-of-the-art in technological terms, and radical in their deployment. The city should be seen as a great opportunity, and it will depend more on us than others who continue to see us in the same light. We don't say it for the sake of it: developing a good strategy for attracting talent on a metropolitan level, investments in line with the productive city model that citizens prefer, self-sufficient and with zero emissions, is all fundamental. It is for the city, but also for world society as a whole. This intelligent city, the Smart City of which everyone in the world talks, where there's myriad companies and institutions that are Smart, but where only us, the ones responsible for the daily improvement of the democratic institutions, are the real City.

Crisis conditions us and it conditions everything. The worst crisis is one of unfulfilled promises, of expectations that aren't met. Besides, in cities we're victims of our own errors and also victims of the unjust fiscal treatment that states impose on a large number of world cities. Nowadays we know we have a path, the path of the internet and the digital revolution, that opens up possibilities with regard to data and the necessary technology to go about making better cities, better solving daily problems, the structural challenges and those of the imperfect market. The more people that use the internet, the better it gets. The DNA of the city, shaped by thousands of millions of pieces of data, that latent genome, so unknown, so trivialised, so difficult to decipher, has as much potential as the project to decipher the human genome. Our goal is precisely to open a door to understanding how to use this genome to better understand the city, and improve it according to individual and collective interests. Will we achieve it? We have to try.

THE AIM: INTERNATIONAL EXPOSURE. BUT WHAT FOR?

Barcelona is one of the queens of individual and collective empowerment, it is in continuous dialogue with that which belongs to everyone; the queen of contraction and determination. However I don't want to lose myself in excessive love of my city. Pune, Nagpur, Cúcuta, Esmeraldas, cities which don't figure on the map of the most well known, are also queens. Just as London, Hong Kong and Istanbul are. Barcelona, and these cities, among many others, is a market, a factory and place to live and a place to see. Barcelona is a community of communities, a neighbourhood of neighbours, a plaza of many plazas. Barcelona has, in addition, a vocation that historically has transcended it, a vocation of cultural, economic and national capital that has not been rewarded by history to the extent of the potential that Catalans have given it. Despite everything, Catalanism hasn't always been aware of the fact that Catalonia wouldn't exist without Barcelona, nor would Barcelona without Catalonia. What we're talking about has not been built without conflict: the capital has become too large to be easily assimilated by an excessively small Catalan administration. The institutional balances after the Transition, and the poor fit of Spanish political and cultural reality, led us to the kind of monstrosity in which Barcelona's potential just doesn't really fit.

I'm referring to the three big challenges facing the Western World: the democratic challenge, the socioeconomic challenge and the ecological challenge. They're challenges that are shared among all modern societies, challenges that question us, challenges to which we need to dedicate more than speeches, we need to confront them with courage and action. Cities have to decide which world they want to participate in and how they want to participate. Barcelona's cosmopolitisation is not neutral: Barcelona's DNA is cosmopolitan. Barcelona's potential lies in its cosmopolitisation; without that Semitic spirit of belonging to itself, at the same time as belonging on a universal level, Barcelona would lose all its meaning, all its interest. Barcelona is continually debating between different ideologies, given that logically not all political options confront global challenges in a similar way.

In Barcelona everything is mixed together. It's a city that is incredibly demanding of itself. They've called it the city of architects, due to the influence of the various schools. It's the city of the industrial revolution. It's the city of beaches and tourism. But it's also the city of neighbourhood associations, which, having fought for democracy, passively allow the increasing questioning of representative democracy: it can be seen in those libertarian "occupy" movements that are based on the nostalgia of anarchism from the 1920s and 30s. Self-management, squatting, horizontal collectivism, are systematically declared antagonistic to the model that has ruled the city for more than thirty years and on which its recovery has been based. The toughest account of the city's struggle against itself is that it stifles its extremities, which opposes investment in and growth of social development. A narrative which argues the first concepts are predators of the second.

We don't agree. We're more interested in the unstable balance between the two extremes: a city that channels its energy to find development models that foster private initiatives committed to equitable and decentralised growth, that generates resources that are distributed in every corner of the city.

A Barcelona that doesn't swallow itself up, nor the territory that serves it, involves devising projects that advance it towards energy self-sufficiency in a few years, without having to renounce any of the advances that the industrial revolution has brought us. The best version of Barcelona is the one that debates, agrees and finally takes decisions and carries them out without prejudices. The brave Barcelona is the one that, as well as risking and daring to find solutions, thinks about precedents that could attract other world cities that know it is not perfect, but are fully committed to all the consequences of its actions. The abundance of contradictions that make up our city have had a very heavy impact on the development of proposals that work to update and improve it. We've seen it in the attacks on big data, Smart Cities and technology in general; this is directly connected to a part of the city's fear of liking itself. People can disagree, be critical, can act politically so that the private sector acts in the most responsible way possible, but despising high value-added economic activity and the innovation sector is literally an error. Because what Barcelona needed a few years ago, and what it still desperately needs, is to radically commit to the diversification of its economy, to favour the development of sustainable innovation projects and, above all, to stimulate the reindustrialisation of the city. Barcelona has a very high quality of life, but it also needs to indefatigably guarantee the right to work. And I insist: along with Barcelona, all the cities in the world.

We've talked about the democratic challenge, and even about the use of the concept to counter the Smart movement. It's certain that the challenge facing Barcelona and all cities remains the democratic one, the challenge of consolidating the democratic legitimacy of our open, secular and free societies . . . that essentially means guaranteeing the social and economic growth of the city. The crisis facing liberal-minded democracies is the crisis of dreaming of social stability, today and in two generations. Barcelona has a young democracy, built on hard struggles and confrontation. The various manifestations of political Catalanism, the syndicates, the movements of a Catalanist and progressive church, cultural and excursionist organisations, the choral movement and the residents' associations joined forces to reclaim the city's resources. Civic facilities, quality public space and inclusive education exist thanks to our parents. One wonders whether forty years later we have the right to say that we are still fighting amongst each other, this time good against bad, or whether we must strive for a deeper level of debate on the issues, well-argued and with a solid knowledge of the causes we are fighting for. Smart Cities and the digital revolution, which lead to a new social revolution that generates opportunities, can be the answer to those challenges. We decided to face up to it without fear, without prejudice, despite not having what other cities in the world can rely on: tools to influence.

Whilst the export agencies of many European and Anglo-Saxon countries have decided to contribute to sustainable urban development across the world (in many cases by utilising cooperation funds as bait to obtain construction contracts), in Barcelona, despite the invaluable talent of hundreds of our professionals, we have found it difficult to work outside Catalonia. After the crisis, the Architects' Association of Catalonia, the FAD,[3] construction companies and developers asked the city council to share costs and join forces to meet the real demand for taking Barcelona's "way of designing" to other parts of the world. It was at this time when the Barcelona Urban Cluster was created to do exactly this.

Barcelona has contributed to organising large events in other cities thanks to the solvency gained after the '92 Olympic Games, and the capacity of its professionals and businesses. The genuine challenge, however, is exposing its virtues, its values, its talent and the way of building cities that we have developed generation after generation, and has been shown to be a good way of doing things. Export the new vision, the new mantra: a city that's energy self-sufficient, moves at human speed, productive in all areas, equitable, hyper-connected, cosmopolitan, metropolitan, with zero emissions . . . The prestige of the Universität Pompeu Fabra and the School of Architecture attracts hundreds of students every year, above all from Latin America, but in a few years we will see it consolidated as a global centre for students, with a higher number of Indian and Asian students. All of them should

---

3 El Foment de les Arts i del Disseny, [Fostering Arts and Design]. A private, independent and not-for-profit association that promotes design and architecture.

come to study, but above all to live Barcelona as an urban laboratory and experience all the areas that have determined its social development and progress in its recent history.

In Barcelona we know how to build a city, now the challenge is to make it affordable and exportable to the whole world. We need this way of building cities, *á la barcelonesa*, to reach the four corners of the globe. We need a change in the local mentality that will lead us to systematically develop the values of the democratic, entrepreneurial, technological, social and open city of Barcelona, on a worldwide scale. A Barcelona with a plan, a Barcelona with a mantra.

4

Opening the City Up for Works

THE TRANSFORMATION OF THE PLAÇA DE LES GLÒRIES CATALANES AS A PARADIGM

The Smart stage is preliminary to the technological stage. The Smart stage involves using the available data to make sure that the change is positive for everyone. The Smart stage involves thinking about how to physically transform the city, whilst providing the maximum possible social value with the change. The data and its cross-analysis can provide many answers that have been hidden up till now. The use of artificial intelligence to shape public space and urban fabrics is already a reality. This is possible through the use of evolutionary algorithms based on the processes of living organisms. With this data and spirit we decided to make the Plaça de les Glòries Catalanes a real centre-point, capable of competing with the Plaça Catalunya, picking up where Cerdà's idea left off and rebalancing Barcelona as polycentric. We demolished the most important road junction and communications hub in the interior of the city thanks to the approval of both the municipal council and an agreement with neighbourhood representatives. As difficult as it was to achieve and maintain the agreement, we maintained it and strengthened it. It would have been impossible to transform that scalextric set into a city if there hadn't been an alliance between the political, technical and neighbourhood authorities. This is one of the important lessons we learned.

The Plaça de les Glòries symbolises the possibility of advancing democratic processes without sacrificing an iota of regulated democracy, without renouncing an iota of the neighbours' committee's capacity to exert pressure and without putting technical excellence at stake. Glòries' transformation was one the great projects during my time at the Barcelona City Council. I knew how to create a highly motivated team in order to address the issue. It was directed by Albert

Civit, head of the Urban Habitat unit and my right-hand man, and complemented by Vicente Guallart, the city's chief architect. I surrounded myself with people from all kinds of professional disciplines, working with the foremost municipal infrastructure company, BIMSA, as the executive arm receiving orders from the immense professional that is Angel Sánchez, gathering opinions and data from all over.

Maria Sisternas, our then young urban projects manager, was put in charge of the transformation of the square. The core of the project, run from the very top of the political ranks, was composed of a cross-disciplinary team of ten people: engineers, architects, transport experts, sociologists, political scientists and economists. On the day of the inauguration of the provisional development of the Plaça de les Glòries we saw Maria's eldest daughter with a balloon emblazoned with the logo of one of the opposition political parties. María and I looked and laughed. The balloon made us realise that the fact that the opposition was campaigning in Glòries meant that they too wanted to capitalise on part of the success: there was no way back now, Glòries Catalanes had undoubtedly started to become a new point of reference for the city, even for our political adversaries.

But how did we get to this point? Imagining how a new city centre should be is not enough; to make it a reality people must systematically use it as such. Centrality cannot be imposed, at most we can induce it, and we dedicated ourselves to doing this: to induce centrality, to *empower* the site by opening it for works. We had started the mandate by displaying some images and concepts developed by Vicente Guallart, architect-in-chief of Barcelona. We gave added prominence to the idea of centrality that would be conducive to human use of the square. This was very different in comparison with what had been designed up till that point. We had entrusted Guallart with studying the possibility of converting the large deserted plaza, planned as a ten-metre high pyramid, into a space for intercommunication between neighbouring districts. A restored space whose most representative elements could be used and enjoyed by citizens. Designing a plaza dedicated to Catalonia's glories and then leaving it without any, means that it has become an irrelevant space, it didn't make sense.

CENTRALITY AS A CONSEQUENCE

The centrality of a place is always a consequence, never a cause. The history of the Plaça de les Glòries Catalanes, will not end with the construction of the Gran Parc {Great Park}. Fifty years ago Glòries Catalanes did not exist as a centre, and it will be fifty more until it fully exists. In fact, all relevant urban transformation transcends mayors and even entire generations of inhabitants. How many mayors of Barcelona can say that they built Eixample? Is Eixample a project or a process? What is more important, the formal result of Cerdà's plan or that it has become the central district of the city's vitality? The architect Albert Serratosa, who said more seriously than in jest that outside of Eixample the only thing that existed was chaos, explained to me on many occasions the reasons for this unique vitality: fractality, flexibility, adaptability, isotropy, amalgamation, rationality, order . . . Could we think in the same way, to use these principles for Glòries?

Our project for Glòries started with the courage to face demolishing the ring road that in theory would solve the traffic in the centre of the city. It was a brave political decision, executed with technical solvency by the municipality-owned company BIMSA, under the expert management of Angel Sánchez, skilfully negotiated and carried out through public debate, transparency and technical excellence. I later reinforced management with the incorporation of an intelligent, sensitive and determined expert in civic engagement and participation, Laia Torras, and premeditatedly integrated her in the project's team. The result was magnificent for morale, and for getting everyone on the same wavelength with regard to the objectives I put forward. The critics will say that the mayor had no choice, that those who forced the decision to tear down the ring road were the neighbours' committee and the previous government's commitment to Glòries. But those of us who thought out and directed the operation, and the group of engineers and architects who worked to ensure the smooth flow of traffic without the need for a flyover, know that over forty technical, budgetary and all kinds of reasons could have been put forward to avoid carrying out works of such magnitude. In turn, politically, in the context of a deep social and economic crisis, the easiest thing was to say that it wasn't a priority, that there were more pressing needs. It wasn't true: Glòries Catalanes is the symbol of how a plan as powerful as the Eixample can get stuck. The symbol of how the most visionary urban planners end up running a communications hub for traffic in the city centre. And how what was designed as a solution, sacrificed the possibility of building high-quality urban space for shared use for many generations. We were not willing to accept it. The Glòries project was one of our priorities for adding new value to the city. In addition, I wanted to do it our way, that is, establishing genuine, critical dialogue with residents of the area, and opening it up to national and international intelligence. Smart plans are not at odds with being open and democratic, in fact quite the contrary. This should be remembered.

The Plaça de les Glòries before the demolition of the ring road and during the construction of its large facilities: The Barcelona Museum of Design and Nous Encants. In the photo you can see how the old plan, designed for cars, forced pedestrians out, disconnected neighbourhoods and created urban vacuums. (Google Earth)

As I was saying, it was a process that was negotiated with political skill, despite having a minority in the municipal government. In private, many people from our political ranks and the opposition said that we had gone crazy, that I "would leave in a hot-air balloon" due to the foreseeable traffic jams in Gran Via, that nobody would appreciate it but that, "it was up to us" . . . Many believed that the works weren't going to go well and it would mean the political suicide of their proponents. We left the meeting in which we finally decided to demolish the Plaça de les Glòries ring road with a new worry: avoiding turning Glòries into a construction waste ground. It was here that we decided to open the city up for works. It is one of the decisions we are proudest of. Thanks to this decision, the small and heroic north Catalan town, Prats de Molló, became the subject of the first dignified outdoor exhibition about the retreat and reception of the defeated Republicans in 1939, in the heart of the works.

The new square's key point: the connection between both sides of the Meridiana, meant a connection between Clot and Poble Nou. This is one of the most important operations that has been carried out in recent years: opening to connect. This is what makes Glòries the new centre of the country's capital. Thus Barcelona served as a the capital even in the trenches of its Smart reform. Immediately the swings and skating rinks arrived which coexisted with the city's most spectacular works. From here we also left with some important political lessons: decisions and their possible negative consequences are taken and paid for in isolation.

It's difficult to understand the complexity of managing diverse aspirations that are legitimate but at the same time contradictory if you don't keep on top of day-to-day activities. We relied on some expert media. It's a fundamental part of the democratic guarantees our society is equipped with. In the case of Glòries only a few people explained what was essential: that city planning measures did not change; that current planning would continue to respect the hectares of grass, the balance of zones and systems, and the buildings that would be created. What can we say has changed about Glòries? In the first place, the aspiration to create a new centre based on the idea of the Plaça de les Glòries Catalanes as a pivoting point and not an end. My conviction is that Glòries is not an urban development problem. Glòries is place defined by the geometry of Barcelona's main arteries drawn by Cerdà, that has undergone changes in form that are closely linked to the urban philosophy of each historical moment, and that had to end this problem by taking advantage of the sign of the times, by taking advantage of our mantra: restoration, human scale, identity, connectivity, technology, equity . . .

THE STORY OF A TROUBLED PLAZA THAT WILL FINALLY GET TO BE ONE

Vicente Guallart, in one of his cheery phrases, told me one day that Cerdà was a genius who, the day he had to draw Glòries, had a bad night. In 1859 Cerdà drew a Plaça de les Glòries Catalanes focussed on the railway (the new invention capable of transporting masses of people in a short time). This passed through middle of the square in a trench until midway through the twentieth century. Coinciding with the expansion of the city, the mass construction of housing and the increasing number of families that owned cars, the General Metropolitan Plan of 1976 turned Glòries Catalanes into a ring road, with intersecting branches of motorway on various levels, seeing things from a vehicle's point of view and perceiving Glòries Catalanes as one of the city's impasses. The Olympics and the urban planning of the eighties unveiled the Olympic Village and then Poblenou, but a traffic-based rationale prohibited looking at the mid-term future with ambition. The Diagonal was extended to the sea only seven years after the construction of the viaduct. This resulted in Glòries Catalanes losing its role as a dead end and becoming a key space for the city. The extension of the Diagonal connected Poblenou to the city, although from 1999 until 2013 the viaduct blocked this continuity. The ring road wasn't demolished to gain only a park, in principal it was done to recover the continuity of two of the city's main arterial roads: Diagonal and Meridiana.

The ring road built in 1992 was constructed using the same logic employed by many developing cities with rampant levels of vehicle use that build them today. It was a question of dividing traffic: an elevated viaduct and a lower distributing ring, with a car park designed to be a drop off point, so that those who came from outside the city could park and go shopping in Plaça Catalunya on the metro. The surrounding neighbours' associations, perhaps with longer-term vision, or perhaps determined not to live next to a ring road, and the sufficient intuition to imagine an open space between all of the neighbourhoods, set up a Monitoring Commission that from 2003 openly pursued the demolition of the viaduct. However, finding the mutual understanding to approve the demolition was much harder than we thought. It seemed like common sense, but people resisted the change. In "the city of architects," the demolition of infrastructure has been anathema for many years, and more so when the design of the infrastructure and the Museum of Design bear the names of people that are relevant to the city: Busquets, Acebillo, Bohigas, Rui-Wamba, etc. They were all generous with our vision: Oriol Bohigas, warm and respectful to me, perhaps because he is one of the most faithful readers of my novels; Rui-Wamba, who designed the demolition project, with the brilliant idea of saving a few pillars for the memory of the place. Many of the critical voices compared the Gloriès viaduct from the nineties to the arcades of New York's High Line, that cross the Meatpacking District (emulated today by the Rambla de Sants, whose name I definitively proposed and ultimately everyone has acknowledged it as their own). They argued that a park should be built above and the space below should be dedicated for the use of temporary markets. Critics of the demolition accused politicians of kowtowing to neighbours' associations, obviating the fact that what allowed the "city of skyscrapers" to maintain the High Line's roads was precisely an unwavering civic and neighbourhood movement, together with a massive private budget to build and maintain the "coolest" park in the city, pending revaluations of the surrounding buildings. The demolition and subsequent restructuring did not make an impression on any other sector.

The most commonly used arguments against the change were the effects it would have on mobility, the cost of the works, the uncertainty over building a new centre, the need to make a paved plaza as opposed to a park and a supposedly poorly drawn-up plan, with too little economic activity and a lack of density. These last two arguments were particularly weak: if 90,000 m^2 of economic activity is considered too little, what was the right amount? If more housing was needed, how much of the area should be allocated to it? 90,000 m^2 is too much to execute in a short amount of time, but it won't be enough in fifty years time. It's impossible to imagine how the square will evolve, but ideally it should densify as its centrality is consolidated. Zoning and metropolitan planning are too rigid. They are in Barcelona and they are across the world. Today we know that zoning a city breaks it on the inside. If in 2007 we approved a plan that definitely won't be executed for a few years, do we need to be so precise with the figures? Incremental city planning should work with quotes, with hypotheses and leave space to be able to increase the density of a place as citizens use them and make them theirs.

Citizens' appropriation of Plaça de les Glòries Catalanes provisional urbanisation, using materials from the works, which are being carried out barely a few metres away.

THE COURAGE TO OPEN UP FOR WORKS . . . IN ORDER TO OPEN UP THE CITY

Open, open, open, that's the key. This is true for Barcelona and for any other city in the world. Eight years go was it necessary to say exactly how a place would be in another fifteen years? For this reason we took the strategic decision to concentrate the developments in multi-storey buildings and free space on the ground for the park, residential use and future third-parties. The political mandate that we had was to turn Glòries into a open meeting place, a link between Clot and Fort Pienc; between the Sagrada Familia and Llacuna in Poblenou. But I knew that we only had four years, in the middle of a transformation process that would last a minimum of twenty more. The challenge, then, was to take decisions that were sufficiently generous, that could be continued, whoever it was that came after us, the development of a reference point in the city without taking any steps that could put the future of the plaza at risk.

In the first place we had to "resew" Ensanche: the railway had altered the course of many of the streets around Glòries Catalanes and our first interventions in 2012 reestablished the continuity of some streets that had ended up as dead ends. In accordance with the Department of Transport I suggested that it was easier to get around Barcelona with the works than without them.

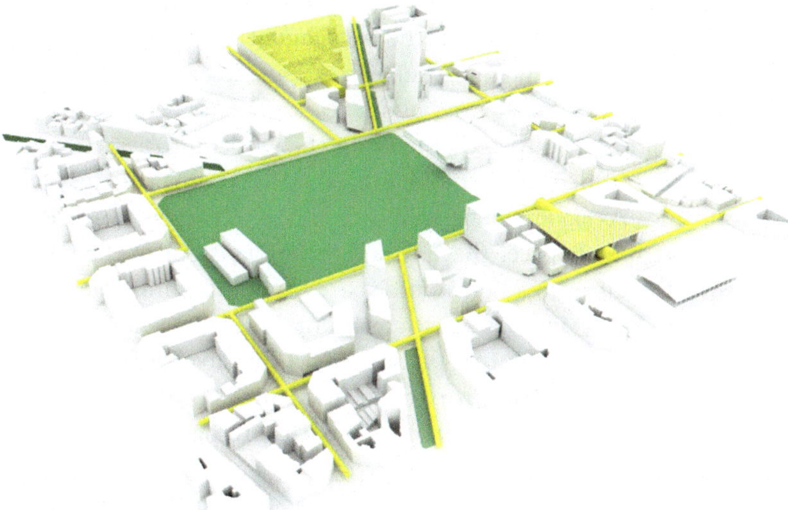

More than 12 hectares were freed up, with temporary traffic passing through Gran Via waiting for the tunnels to be built, which started in April 2015. I wanted to dramatise the change in the transportation model, and where before there was a loop with three lanes for cars and two for buses, there were now bike lanes, quick lanes for buses, children's games, athletics tracks for running, skating or whatever else. We were lucky that, as we had a logical budget constraint, the spaces that we had to transform were very lightly retouched. The project management acted brilliantly once again: with painting, temporary trees, urban furniture recycled from other parts of the city one of the highest quality public spaces in the world was created. If creating a permanent square in Barcelona had an approximate cost of 350 /m^2, the provisional installations in Glòries had a maximum cost of 60 /m^2, and all of the urban furniture could be reused for other locations.

In accordance with the motto "open up for works," my team made sure to reserve a part of the budget for the works at Glòries in order to marginally improve the space freed-up by the demolition of the viaduct. It wasn't easy to make room in the budget for what we called "provisional development": the item agreed with opposition groups was strictly for the demolition of the ring road. From the start very few understood the need to reserve 5% of the budget for improving the recuperated space. The harshest criticisms spoke of "the ugliest public space in Barcelona," claiming that it was a worthless expense whilst waiting for the permanent construction of the park. We ran the risk of managing an urban desert or "closed for works" building site for at least eight years. The strategy, as I have already highlighted, consisted of making the works into an attractive and educative public space: colonising every free corner of the plaza for public use. As soon as the mobility model changed, we had to change a centripetal space, that pushed pedestrians out of the centre of the circumference, to a permeable space, integrated into the fabric of Eixample. The transition, one of the most complex in the city, required intricate planning in phases. The provisional development was designed by the Urban Projects department, in other words from within the Council, but in collaboration with external teams of architects. Sisternas provided instructions in two clearly distinct parts, which in fact were executed independently and in phases, as mobility was resolved. The development on the seaward side was the first that had to be executed because it was the first part to be freed by the demolition of the ring road. However it wasn't easy to give them civic uses because they are spaces that join large features of the city that rarely communicate with each other.

The provisional development consists of a plot of land in front of the Mercat dels Encants on a steep slope that is being reused as a temporary nursery for the holm oak trees that were planted on the interior of the viaduct. Mar Escala, Adolf Cruces and Carles Casamor designed one of the most celebrated spaces in Glòries: a terraced space connected by gentle slopes, to be able to take care of the oak trees with the heavy machine that is used to move or water them. The slope was taken care of with gabion retaining walls made from crushed concrete from the demolished ring road. By doing so we avoided having to move materials to the collection centre, at the same time as creating an area that has a very refined language, where walls and benches for relaxing are one and the same.

It's an intelligent solution that works well whether there's people there or not. If it's empty it's a working nursery and an incredibly leafy area. If it's full of people sitting on the benches it's a place of comfort that produces shade. A place that foreshadows the intensive use of a green park whenever the Canopia Urbana [Urban Canopy] project, work of Agence Ter and Anna Coello and winner of an international competition, is implemented. Additionally, it's a space that is ordered in the grid style of Eixample and that some people cross on their route from the Barcelona Museum of Design to the National Theatre of Catalonia or the L'Auditori music hall.

However, the most important thing about this project, without a shadow of a doubt, was knowing where was the best place to put pedestrian crossings, favouring the flow of people and making circulation around the plot comfortable. Opening up, here, turned into connecting neighbourhoods, opening routes, letting people from both sides of the plaza start to use it as one, to make sure they felt it was a connector, not a separator.

We were obsessed with making a space that worked in a circular way with the roundabout and also had the same orientation as Eixample. I asked project management to stimulate local talent. Maria Sisternas held a mini competition among highly regarded young architects to build three temporary constructions that would help to mentally place people in the context of the orthogonal grid. One is the Visitors' Information Centre, by Peris-Toral architects, it's a passageway style building which is extremely well executed and faithful to its original idea. It's a very simple wooden box, with a translucent roof and walls that give it brightness. The box has an expressionist cover, like a loose, flowing shirt that moves over the body. The unique feature of this pavilion is that it can be walked through longitudinally and transversally, with a very simple and intelligent door system, which therefore helps Glòries to be a place that can be explored on foot.

From day one the city responded positively. The city's verdict on what we were doing was clearly favourable: the new plaza was used from the very first day. The project worked perfectly, inviting the immediate appropriation of the square. The mix of the residents' self-adjustment and the good work of the city's technicians, and young talent turned the Glòries project into an international benchmark. Opening for works, giving private mobility its fair share of space, relying on the collaboration of creative people from all fields and of all ages to invigorate a space that still wasn't finished. We brought dozens

La Plaça de les Glòries, in the middle of its provisional urban development, with the connections to Eixample open, the recovered space for neighbours and the transcendental connection on foot between the two sides of the Meridiana, where the historic exhibition on Prats de Molló and the Retreat took place in 1939. (Google Earth)

of urban leaders from all over the world to walk through the plaza, and everyone considered the achievement a success. The conclusion is obvious: there's no turning back now. Glòries is already a vital new area in the city. By opening up the area under construction for continued local use, we blurred the barriers that cause popular dissatisfaction whilst regeneration work is ongoing. But we did something else: we provoked the neighbourhood's appropriation of the available spaces with a dual purpose: in the first place, to avoid the social desertion of this urban area; secondly, to learn how to use these emerging spaces by interpreting them in the same way as those who, in time, would be their only protagonists: the neighbours.

What was most important was to implement the dream that up to this point hadn't be accomplished: to open the city up for works. Open, open, open. Practically from the first day we announced the demolition of the viaduct we decided that the city and the residents should view the works in Glòries Catalanes as an opportunity. For us it was a fundamental operation, almost ethical: to destroy a ring road that was so important was quite something, and the available transport studies, no matter how much time and money we had invested in the reconfiguration of Eixample's surroundings that were still closed, did not give us certainty over what was going to happen.

SOME USEFUL LESSONS

The lesson that Glòries Catalanes taught us is that current societal changes are produced at a vertiginous rate. Thanks to technological advances and the digital revolution in the last fifteen years staggering changes have been produced in all aspects of daily life. Fifty years ago urbanism could afford to plan thirty years ahead, without having to change anything in its plans. Today, change is tangible in less than a political mandate, and the design of a city should not be unaffiliated with it. Flexibility is imperative and the best professionals are those who can understand and project accordingly. Just fifteen years ago central areas were designed according to this blueprint: build a thou-

The new Plaça de les Glòries has a central, connecting, restorative, symbolic and appreciative role that opens new perspectives on how to create public space around the world.

sand houses, install some facilities on a city-wide level, connect to the metropolitan area, carry out a communications campaign, invite renowned architects to design housing packages and public space, install the offices of a multinational, do everything with views of the sea and give incentives so that all types of liberal professionals buy houses. Ah, without forgetting the iconic buildings signed by world famous architects.

If this was the formula before the crisis, things have changed since the fall of Lehman Brothers. The Urbanist's Manual has expired, in part because of the resounding explosion of the housing bubble, and in part because, fortunately, more and more professionals in the city speak the language of neighbours and citizens, and citizens and neighbours listen to, read and personally know architects and city planners. The recent Pritzker Architecture Prize awarded to the firm RCR from Olot, Girona is largely proof of this. Today, questions like comfort, quality of life, human scale or proximity have gained relevance – the transformation process is as important as the final result. I wish that the transformation of the Plaça de les Glòries Catalanes had not started with a 158 million euro investment in two installations like the Mercat dels Encants or the Disseny Hub. I would have started with public space, but there was no way back. The two buildings have obvious virtues. We decided therefore to work with the more positive aspects that they offered to the city. The Mercat dels Encants (a flea market), which gave me so many scares before its inauguration because nobody took into account its potential as a rainwater accumulator, should have worked as a ground floor like all the flea-markets in the world. However, the plot of land it was assigned was small, and the competition's rules solved the problem by including a spiral street-ramp. We believe that in 2007 maintaining the Encants de Glòries' activity was a success; the number of people present when during market's opening hours is nothing short of remarkable.

Skeptics claim that what has happened in Encants is full-fledged gentrification, that fewer antiques and junk are being sold and that now we are starting to find products for tourists. This could well be true, but the new market is a resounding public and economic success, in addition to ensuring the maintenance and creation of new jobs. As long as this activity remains and you can find bargains, for example a painting by Ràfols Casamada, which appeared at the end of July 2015, then the transformation is welcome. Without doubt there will be another Encants in some other place on the outskirts of the city, and perhaps in thirty years, when the development has been paid off, we may well decide to change the purpose of such a strategic piece of land. In the meantime, we have to work so that the intensity that is felt in the Mercat dels Encants spreads across the whole square. So that the spaces' boundaries become blurred, attracting people to this unique space, guaranteeing an influx of thousands of people to turn it into a meeting place in the midst of anonymity.

One of Glòries' challenges is to make sure that the uses of the space on the plaza's fringes complement each other. Have you ever tried to have dinner near the Auditori or the National Theatre after an evening function? It's almost impossible . . . The Auditori and the National Theatre are two buildings that are clearly closed off from everything else, they don't generate any kind of relationship with their surroundings. The Plaça de les Arts is one of the outstanding issues to do with Glòries Catalanes, but also the city as a whole. It's logical because at the time it was built (1990–99), Plaça de les Glòries was not a place, it was a motorway exit, and the Diagonal didn't even reach the sea. Let's think positively however, and remember that it was Councillor Max Cahner's vision that prompted the National Theatre to be built in what is today becoming one the most important centres of the city. The city should be able to cope.

However, apart from the Museum of Design and Encants, there are another fifteen facilities that need to be built in Glòries, if we're being strictly faithful to the political commitment of 2007. In the coming years we should rethink the programmes of developments that are going to give the park a façade. It would be ideal if they were developments on a local scale, as reflected in the commitment, because this will ensure that the edges of the park are full of people. This is why we decided to build the new school of Encants on the corner of Carrer de Cartagena, and as such we left everything ready to build a nursery and a visitors' centre. This corner will become a hive of activity that will fill the park with adults and children on a daily basis.

The new developments in Glòries, on a neighbourhood level, will have to rely on programmes that guarantee the diversity and influx of users with outstanding architecture. Furthermore if, as we plan, we are able to renovate the Rec Comtal, a medieval irrigation canal, and the Fábrica dels Paraigües by Pio Rubert Laporta, a twentieth-century umbrella factory, Barcelona's history will be better appreciated and it will be understood as the urban palimpsest that it is. Some designs are stubborn, and Glòries' should contribute to recovering the pieces of urbanism that Ensanche, in spite of everything, could not erase.

The project whose motto is "Urban Canopy" that won the international competition is a brave investment. It's a project aligned with Barcelona's Smart mantra that relies on restoration, trees, and large pockets of greenery to create comfortable areas in the thick of the urban vortex. It's a park with mediterranean vegetation that creates shade and distinct zones, with different uses and something for everybody, designed for activity twenty-four hours a day. The project recuperates Diagonal and Meridiana's continuity, leaving a large agora for varying uses as a unique and flexible space in Barcelona: a large grass court surrounded by walking paths. But, above all, it's an innovative project, that doesn't depend on architecture to create centrality and whose protagonist is green space.

The most symbolic part of the plaza still has to be resolved, the artistic element that will imbue it with the collective identity equal to its name and location within the city. It is convenient to work with sculptural elements, a piece that is visible from the rest of the city and that recognises the centrality of the three avenues' meeting point. Why not one of Jaume Plensa's challenging effigies? It could be a great opportunity to exhibit the work of one of Barcelona's great sculptors who has experience in working with public space. In fact we tried to put one of his effigies on Barceloneta beach to greet the world. We looked everywhere for sponsors. Some were immediately available. The usual ones. Many times the most criticised ones. They tend to be the most envied. The others, also the usual ones, raised their eyebrows condescendingly.

The strategy applied to Glòries offered us a 21st century rereading of the composition of an agora, that point of the city that heralds it as a genuine polis. Let's see why: agoras expanded their role in classical and Hellenistic times, becoming the manifestation of civic order and social life, with their civic and religious temples, money changing tables, incessant trading of oil, spices and all kinds of goods. Agora, therefore, was logically synonymous with collective belonging, and social and economic appreciation. And along with the agoras, the markets, the mediterranean-style lodges, the best of which by the way were in the Catalan capitals Valencia, Palma, Perpignan and Barcelona. These agoras, together with the lodges, and the porches, were authentic platforms for civic interchange, in which it was possible to find Saul of Tarsus arguing with his fellow philosophers about the idea of one God, or Michelangelo waiting patiently to discover where his marvellous statue of a hesitant David would be placed.

Beyond this, the agora was also the manifestation of justice, the point at which the transparency of civic or economic transactions was judged. As these and other deliberations were moved to noble buildings, the agora became a forum: a place to stroll and exhibit, to see and be seen, a stage and its stalls, a place where individual identity was shown and collective identity was felt. The forum, the evolution of the agora, in essence the agora transmuted, was the place where we grew collectively, as demonstrated not so many years ago by the revolutionary movements that occupied the best squares in the world at the start of our century. Zuccotti, Catalunya, Tahrir, Taksim, with their rundown shops and idealistic-apocalyptic discourses, for me were nothing other than the vindication of the square as the place in which we can all say what we want, the place that belongs to us, the place that is ours before anyone can decide otherwise.

Opening up for works opened our eyes to a definitive reality: people look for the sun's embrace, the gentleness of the shade, a space for face-to-face conversation. In our Mediterranean countries (and in those that aren't), people, like Aristotle, know that we can call a place *polis* where a herald, a crier can launch his tirade and be heard, a place where we can see faces and enjoy conversation.

The idea was not to short-circuit the ecological dimension of the city, the dimension that ties us to our surroundings in a definitive way, no matter what happens within them. Great transformations have to be experienced daily by their neighbours, by the whole city. And the works have to be envisaged with sufficient intelligence in order to start leaving the embers of civic appropriation that guarantee a good transition for the city. In this way, the transformed plaza, street, and market will belong more than ever to their protagonists. The people will perceive this new public space as pertaining to the people, it will be given the function that was planned, a space for shared identity, for the shared experience that we want for our cities. Because the best squares, the best places in a city, are a state of mind.

The night the jury issued the verdict, Joan Ixaso, a tough neighbourhood representative, stood up from the table to give thanks for the transparency of the process. Then, without warning, he sang

"Maitetxu mía" for the group of privileged people among whom I found myself in the large dining room of 7 Portes (a fashionable restaurant, which first opened in 1836). Now there were no "sides" between politicians, experts and neighbours.

URBAN TRANSFORMATION AND MANAGEMENT OF SOCIAL COST

The works on the Rambla del Sants, which coincided with the ones on the market and the surrounding area, caused enormous damage to the neighbourhood. The final result is spectacular. However, had they thought about opening that part of the city up for works, the pain involved would not only have been mitigated, they would have also have generated the interest and appropriation that the new space deserved, and deserves. I had discussed it with the headteachers of the nearby schools. One of them told me that the city had become invisible there. And with it, the school. And with the school, its students, families, teachers, administrative staff . . .

The transformation of Barcelona's Santa Caterina market appears in all the touristic and architectural tours of the city, but it had a high social cost during its execution. Santa Caterina was one of those places where Barcelona still lived with its past. Under the colourful ceramic roof a lot of lifestyles have disappeared, of which only a few bars were recovered. The market resisted as best it could, but the neighbourhood disappeared. The works lasted eight years, during which almost no-one had the courage to speak about the disaster that submitting a neighbourhood to a generation of urban mutation implicates. My grandfather had been the manager of a coal-yard in front of the door to the market, in what today is the Avinguda Cambó. The market was run-down, full of people, activity, staff who had come from those battered, dingy and foul-smelling ancient streets, but it was full of what makes a real city: different lives intertwined, interwoven, overlapping. My grandfather's coal-yard was one of the neighbourhood's meeting places, given the man's social capability, always ready to abandon his workplace for a fine-cut cigarette and a glass of paint-stripper-like cognac. The coal-yard wasn't a success; some time later he had to leave it to open another one on Carrer de Cruz de los Canteros, in the middle of Poble Sec. But its story remains, between the stones of the Santa Caterina neighbourhood, just off the noble and distinguished calle de Mercaders. Cognac, clamour, old decrepit buildings hiding bourgeois and naval elegance, today forgotten. The market, the fruit of expropriation, executioner of the "so they say" most beautiful gothic monastery of medieval Barcelona, the agora of all agoras. Between Santa Caterina and the Sant Pere neighbourhood, between the market and what today is the Pou de la Figuera, for three centuries Barcelona has been struggling to recover from the disaster of the destruction of the Ribera neighbourhood during the War of Succession (1701–14). It has been half-achieved, because in that part of the city one has always felt smaller. Despite everything, people from Barcelona never fail to revisit what we consider the true heart of our citizenship. We were *ciutadans* long before the French Revolution started to hand out citizenship cards. And we are *ciutadans* because in those market streets a commercial empire was founded in whose remains we still live. When the controversial and sometimes visionary politician, Francesc Cambó created the separation for the Vía Layetana, Barcelona started to heal through that *embolic* part of the city. The union of the Vía Layetana with the axis of the streets Ferran, Jaume I and Princesa opened up the city in its political and symbolic centre: the Cathedral – Plaça de Sant Jaume axis, or the *cardo maximus* of our seminal roman city. The polis reincarnated from its implosive ashes.

When the political model for the recovery of Barcelona's markets started, during the Olympic years, it focussed on Santa Caterina, a correct decision that was strategic in the scope that was taken. The municipal markets were beacons of community life, and their central and strategic locations in the city's neighbourhoods granted them an incomparable urban-strategic value. Recovering the markets through their architectural regeneration, coming to agreements with retail area managers, creating partnerships with the private sector, were wise liberal and social decisions. The centres of European cities have historically been their markets, and rehabilitating them has been a central policy of cities such as Rome, Paris and London. In Barcelona the meticulous operations have been done excellently, like in the Mercat de la Concepció.

The new Santa Caterina market is already over ten years old, and it still looks new. While being located in Ciutat Vella and having all the characteristics to be able to compete with Sant Josep market, it is far from achieving even half of the visitors it could attain. And all this despite its beautiful and elegant design, and beautiful it is, and the municipal efforts to reposition it on the map of Barcelona. The reconversion of Santa Catarina market forced the traumatic reconversion of a neighbourhood. Santa Caterina was a case of time management: closing the neighbourhood for more than six years definitively thwarted the option to reconvert the small businesses that gave life to the arcades, surrounding alleys and the neighbourhood.

This was the origin of our obsession with time management during works. At that time some of us began to forge the idea of collating works and civic appropriation of the spaces subjected to them. A building site is an internal border, a trench in the middle of the city. The mutation of urban space inevitably alters the daily routines of people that cross the city, that go to school, or browse the shops. Works, in general, are tolerated for the supposed mid-term benefit. But in many cases, the damages are irreversible. We proposed to ourselves that we would minimise the impact of the works, manipulate their limits and favour the creation of contact spaces: areas where the city conquers spaces and allows urban life to flourish. The same works are a spectacle: there are always people spying on the fences, watching the movement of machinery and workmen. Intelligent management of urban space has to do with its human and humanising management. Trust people, tell people – open, open, open the city up for works. Do it with the intelligence that the available technology provides us, big data, small data, open data and all the data that comes to mind. Give this data to the city's Smart service, to open it, to connect, to make available, this is the main lesson of Glòries, the rambla de Sants and of Santa Caterina.

5

City-Os, Big Data and High-Resolution Diagnosis

THE CHALLENGE: ANTICIPATING CHALLENGES

The truth can be tough, most of all when it doesn't fit with our idea of how it should be, or when it goes against us. Tackling a city's transformation demands that fine-tuned diagnostic capacity which we sometimes lack, or that sometimes we are afraid to accept. Smart Cities provide us with it as never before, to a level of detail not seen until the present day. The digitalisation of reality allows us to understand people's condition and living conditions on a person by person basis. Furthermore, it allows us to add data like never before to be able to adequately budget and project, without falling short, without going over. If world leaders were aware of this, how could they leave anyone behind? It's the first time in the history of humanity that we can handle the provision of services, the definition of public space, the creation of installations with such precision. What are we waiting for? Reality is profiled for us in high-resolution. Bettering services and transforming the city require addressing all scales, from an individual person to the population as a whole; from a specific flat to all housing in the city. In this way, in high-resolution, the capability to diagnose point by point, the capability to forecast for the population as a whole, or to cross-reference data that allows us to anticipate or to take immediate action, becomes make-or-break.

In Barcelona we wanted a city that was capable of anticipating the great challenges of the 21st century: transport, energy, urban challenges . . . We thought of IT-type tools, heavyweight systems capable of telling us what we consume as a society, how we move. But we also wanted a city that was capable of anticipating the daily problems of Mrs. Carme in Barcelona, or of Mrs Sita in Bombay. We wanted to know if these elderly, lonely women, who could become tragically neglected on a daily basis, are eating every day. We wanted to know if they are healthy, if they are taking the pills the doctor has prescribed them. We wanted to know if they go out for walks, if they have social relationships. We wanted to know if they can pay for electricity and gas. And we want to know it so that what happened in Reus doesn't happen again, because today it's possible to know. The difficulty lies in identifying these challenges before they exist, when they can only be sensed.

As a result of these reflections, we started to focus on new intelligent services, or on overlapping services, rather than on vertical solutions. If we needed to be intelligent, if the city needed to be Smart, it had to answer to a different logic than the one that had been implanted up until that moment, based on the vertical flow charts that we all know. In fact, the creation of Urban Habitat made this task fundamentally easier. Lighting, for example, that is to say public street lighting, stopped being only that, light and that's it. Lighting was immediately seen as a key factor in safety, improving comfort, economic activity and recuperation of identity. Nothing new? Let's see: we immediately started to think about variable types of lighting, connected to security systems integrated in mobile phones. Let me explain: let's suppose that Mrs. Carme's granddaughter Judit, eighteen years old, plays basketball and comes home from training late. Let's suppose that some of the streets she walks through are particularly deserted at that time. And let's suppose that Judit has an alarm button on her phone that allows her to turn on street lights she passes under, either to increase visibility, or to warn the closest police station and her home if something were to happen. If we also have data at our disposal that actively monitors crime in certain parts of the city, and we manage to combine passive prevention with active involvement, both the sensation of an improvement in security and its effective prevention immediately increase. This is a new civic service, in which, by deverticalising the management of public services, using diagnostic capability and combining it with the ability to react immediately, a Smart City reveals itself as a new tool with unknown possibilities until now. Perhaps the installation of sensors in some public parks, which need intervention in the cleanest and least aggressive way. Combining energy self-sufficiency, renovation, and big data allowed us to modernise, to get to know our immediate natural environment better and the way it enhances intelligent leisure amongst our citizens.

Until today it took vast intelligence to able to correctly predict things. And we weren't always right. From this reflection we came up with the idea of a City-OS, the city's operating system. My colleagues Manel Sanromà, Josep Ramon Ferrer and Eduard Martín developed it, uncovering that none of the solutions large companies offered us were fit for our purpose: to identify point by point, to create structured policies, to diagnose, cross-reference verticals, work with all residents, interpret data from the sensors installed across the city . . . Eduard Martín, with his incredible pedagogical quality, explained the City-OS as if it were a Smartphone: a "simple" device that integrates data and applications developed by whomever: government, university, start-ups, NGOs . . . with the ability to inform and to take action.

THE CITY'S OPERATING SYSTEM

Barcelona's City-OS is without doubt one of the most important conceptual advances of the 21st century in the city's operational management, even though it is still in the making. Many very good companies are fighting to develop this tool all over the world. At present it's difficult to see anything of any importance installed; the error still resides in the fact that the majority of mayors, high-ranking officials and public and private specialists only focus on the technological solution. However, the key is in the development of intelligent cross-field services: combining security and lighting; combining gas and electricity consumption with home ownership; combining age with physical activity, training and socioeconomic data; combining the use of public transport with age, daily movement, the price per square metre and victimisation . . . Using new technology should serve to decipher a city's mysteries. That's what the City-OS is for. Because in reality we know very little about how we live, and that's logical, because the complexity of a city increases exponentially. Having the right information available is essential in order to plan an equitable city. The inertia of cities is tremendous, and to guarantee an affordable housing market, for example, you have to begin to coproduce policies with small and large landowners, whilst making correct diagnoses, using accurate data, and coming to practical, judicious conclusions, freeing us from stories fraught with banal prejudices.

In la Oreneta park, Smart connectivity, sensors and energy self-sufficiency.

A city's operation system, or City-OS, is a layer of decoupling between sources of data and solutions for intelligent cities. It allows us to decouple technological solutions from their providers, define the city's complex processes that enable the taking of real-time decisions, whist building an expansible and replicable system. It is comprised of four main elements:

1. An ontology to organise data.
2. A primary data repository.
3. Semantics to unify and globalise access to the data.
4. A "business" process container; in this case, "city processes."

The repository has three main levels of data:

A. The staging area for loading data.
B. A unified archive of standardised data.
C. An area to publish analysed data.

It can be seen in the City-OS scheme developed by Eduard Martín, the leading IT engineer in Barcelona's digital innovation who was able to conceptualise and convert some of our visions to architecture description language:

City-OS is a simple information technology construction based on the three traditional layers of computational architecture: presentation, business and infrastructure. The separation of data sources and infrastructure (Infrastructural Level) from the business level allows the addition of any data source (traditional databases, sensors, actuators, data from carriers and private companies, internet, social networks . . .). This data has to be standardised through a "standard solutions interface". The interfaces between the different layers are the City-OS's "open secret". The separation of the process level, where the "big data" concept really materialises when a layer of related information is created, comprises the brain of the system: database acceleration (data that becomes information), ontology and semantics, machine learning and user control, allow the configuration of complex city processes that combine information from different verticals. Another standard interface allows us to communicate with the presentation or application layers, to build "responsive" systems that enable the presentation of information according to need. Access to the City-OS world is multichannel through the various levels of security applied to every standard interface.

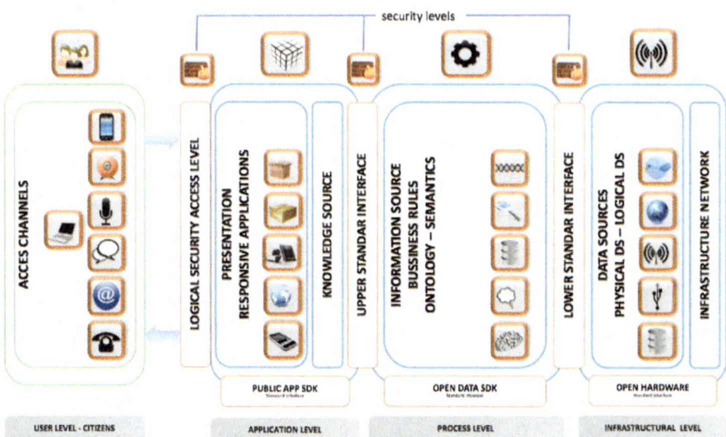

City-OS aims to ensure security through levels, and using the "machine learning" applied to the intermediate layer of processes, make sure that the information coming from the logical data relationship is converted into knowledge for forward planning and retrograde control. The "Open Data" and "Open Hardware" paradigm is structured in their respective layers, achieving safe ordered access to appropriate computational levels.

The City-OS is like a motherboard to which different providers of solutions can connect their modules or intelligent city functions. In turn one city's City-OS can be connected through its standard interface to another City-OS. Every City-OS can be configured for different purposes, whether using open source, additional modules, or additional modules with a license. However, the way we intended it in Barcelona was for the nucleus of the City-OS and its modules to be open source. An operating system can differ from city to city, it can have different functions, but will have the ability to communicate with another through the exchange of information. Can you imagine? So much talk about the need to empower cities, so much dialogue on how to give cities more influence, and soon, thanks to the connectivity of world cities' operating systems, we'll be able to build a genuine global operating system to improve the living conditions of thousands of millions of people. Better services, better public spaces, better quality of life, clear air, safer cities (the "Ciudades en Red," or, "Network of Cities," concept brought to life through computers). The City-OS, in short, is the operational core of Smart Cities and traditional systems of civic management, capable of turning data into high-resolution diagnoses that we must immediately turn into predictive and analytical capacity and the creation of concrete solutions to complex problems. To do so it's important to get back to individual people and the pressing problem of the right to housing, as prevalent in Barcelona as in any other part of the world. It's important to review the situation using real cases.

THE VALUE OF THE CITY-OS LIES IN CIVIC SERVICES

Patricia Belasco is Ecuadorian. She's been living in Barcelona for nine years, in the Zona Franca area of Sants-Montjuïc. Patricia came to Europe with the hope of working for a better future for her and her family. She left, or rather fled, the Andean plateau where she barely earned enough to put food on the table for herself and her eldest child,

Dandy. So much so that Patricia, recognising that she couldn't get by, decided to send the boy to Guayaquil, where her in-laws live, and she left him there. After a while she became pregnant with her second child, Washington. The news, despite seeming positive, looked like a bad omen: a new mouth at home meant having the same problem she had just escaped with the help of her in-laws. Is there anything more difficult for a mother being separated from her child? And is there anything more difficult than separating for lack of food? Patricia had just about managed to do it with Dandy. Now, with Washington, she had to face her fate. What should she do? Ecuador didn't offer her answers. She finally took the advice of one of her friends: she had to follow the route to Europe, specifically to Spain, where things were going well, there was an abundance of work for Latin Americans, and she would possibly be able to get Spanish nationality. For a time she had doubts. After two years she had to make a decision: Guayaquil or Barcelona. She chose Barcelona.

They were eight far harder years than she could have ever imagined: adapting to Barcelona wasn't easy. She discovered that the "homeland" was not as homey as she might have thought. Looking after a child without the support of almost nobody is very difficult. In fact she only had the support of social workers, school teachers and her friends' solidarity, the victims of economic exile. Life, however, takes unexpected turns which can appear at parties, or whilst meeting friends. First it was that dance, then it was that coffee. Eduardo started to enter her life and one day they decided to live together. Eduardo worked in a company subcontracted by another, in turn subcontracted by a developer that built flats in Vallès. 2005, 2006 went by . . . Why don't we buy a flat, Patricia? After all, with your 800 euros and my 950 a month, surely they'll give us a mortgage. In July 2011 Eduardo lost his job. In October 2011 they paid off their debt to the bank with their flat so the bank wouldn't make a claim. The non-recourse debt had worked its charm without anyone noticing. What nobody had talked about was that nonrecourse debt meant renouncing your future.

On the 12th of February 2014, I was sitting on a plane on course for the Americas, Patricia was sat next to me and told me her story. She had to return to Chaco, without Washington who didn't want anything to do with Ecuador, without Eduardo who already had Spanish nationality and didn't want to go back to his country because he would rather die than admit the disaster that had befallen him. Without a future, without her children, without her husband.

The question is obvious: what could the city offer Patricia? What civic services? How could we do this without knowledge of the processes that led her to this situation? She, who contributed what she could to make Barcelona the city that it is, who has formed an active part in the construction of the city, could have, should have hoped for something more from it? The answer is yes. What we have to do is work on a diagnostic level to find a solution for Patricia before it's too late, and to know the exact steps we need to take to cover those needs. Intelligence should be at the service of these challenges.

With data, with a good operating system capable of cross-referencing it, automatic learning algorithms and good cartography we could have a map at our disposal that shows types of property owners, the size of smallholdings and the occupancy of properties. The data could even be superimposed on the investment made and the price of the leases, thus being able to see the trends of affordability in different cities. Inequality can be combatted by making more information available to all stakeholders; in other words, to the general public. We need a tool that designs policies for land acquisition for cities' urban agencies. We need to introduce new tools to stimulate that old aspiration of the urban political agenda, the acquisition of public land for future infrastructure (housing, facilities, research centres) that would counteract some of the dynamics in the market. The private sector is not predatory by nature. There are speculative companies, yes, but the majority buy without wanting to go against anybody, they are in favour of activity that generates more activity. If city governments knew how to access such data and explain the general interest in reserving public plots of land for future generations, it would surely be possible to reach agreements with landowners.

HOUSING, BIG DATA AND SMART CITIES

In Barcelona we already know what the real needs are in terms of housing. The civic housing debate had been based on particularly poor and self-interested diagnoses. It had been made clear that there were 80,000 empty flats in Barcelona, and it had been said that the banks owned thousands of flats they had systematically obtained by evicting people in vulnerable economic and social situations. This account, which is quite accurate for the metropolitan outskirts, is however rather inaccurate for the city of Barcelona. It's a false affirmation, the result of general ineptitude, especially the Council's, to try and understand how we really live, beyond what we publicly declare. The extent of the empty flats phenomenon is ten times less in Barcelona than what some people claimed, but what's most serious is that critical voices were not raised when we proposed that the facts should be corroborated. The projection of 80,000 houses was based on the number of flats where no-one was registered as residing. The truth is that the Council hadn't put the proactive monitoring of housing stock into practice. The diagnosis relied on the residents communicating where they wanted to be registered as living. The 80,000 empty flats were estimated on the back of the number of flats where no-one was registered as living. The reality, however, is far different to the data that the city council handled. For example, more and more people are living in one city but working in another.

Contrary to what had been said for so long, today we know that housing stock in Barcelona is not large enough to dominate the market. We're seeing it again now in 2017. Without 4% or 5% of flats available, inflation is guaranteed. This is the real diagnosis, provided by Smart analysis of the available data, cross-referencing different sources and screening them through the City-OS. Thanks to big data and its management, today we understand the city's genuine needs. And with them, we must work to offer a real alternative to the market's failings. At the same time we must provide online training so that people stop trying to trick us with financial services that are designed to confuse. The city will also have to offer an appraisal system that complements the bank's systems, to ensure that no-one has to pay exorbitant prices for their house. The city must help Patricia to seek advice that gears her abilities towards a business, or offer her services that guarantee her a fair salary in exchange for serving the city.

Questioning the state's legitimacy to drive urban growth rates is the order of the day. In 2008 the Spanish President Rodríguez Zapatero created an extraordinary fund of 11,000 million euros, equivalent to 1.1% of GDP, to create 300,000 jobs and revive the state's economy. Meanwhile, Spain was falling into debt through the construction of infrastructure and facilities of questionable public utility. It would have made more sense to invest this sum in buying land, instead of those useless airports built in Spain in the last years, and high-speed trains in which the state has invested, which do not transport anyone.

In the field of social services improvements can also be palpable. Let's take the case of overcrowded housing rates. There are meagre apartments where dozens of people are registered as living. If this is true then the situation is serious and action needs to be taken to look after these people. If it's not then this creates a double error, because additionally it is counted as an empty flat when in reality it's occupied. The casuistry is diverse and includes dysfunctions from detecting squatters to flats where no-one is registered and that are earmarked as studios or workshops. Innovating in the collection of data on the city is the only hope to better manage access to housing.

According to verified data from the Housing Authority that was obtained by cross-referencing consumption data, social services data, censuses and inspections, exhaustive field work in representative environments and extrapolating it on a citywide-scale, we know that in Barcelona there are around 820,000 housing units. Of these, the incontrovertible use of about 755,000 is for housing and around 27,000 have a different purpose (offices, workshops, studies, etc. . . .). It seems that there are almost 31,000 housing units that are in disuse (approximately 3.8%), of which two-thirds are on the market (waiting to be rented) and a third are truly empty: therefore there are about 10,000 empty flats. Finally there are around 6,800 flats which are being refurbished or are in ruins. In short, the prognosis that that are 80,000 empty flats is completely wrong: there's not too many empty flats in Barcelona, there's too few. When you miss the mark with the prognosis, you miss the mark with policies. That's what the experts say: in the White Paper on Housing, whose publication was coordinated by Professor Anton Costas, it is stipulated that for an adequate residential system we need 5% of housing to be empty as a minimum. With less than 5% of housing stock empty, the market doesn't work.

In Barcelona we need to build more to be able to adequately meet the demand for housing and avoid generating a housing bubble. A housing market with affordable prices for both the public and private sectors (each is a consequence of the other's existence). The data tells us that it takes long-term vision to find that land necessary to build the stock that the city needs. Data and exactitude come to the fore. The new housing model must combat barriers to affordability; it must encourage young people to become independent and engage, it must ensure the variability of activities and residents, it must take in people who are passing through or in emergency situations, it must take care of elderly and disabled people and it must be energy self-sufficient housing during its lifetime. Hence the most important thing is to acquire or create land. The legacy that we will leave for future generations depends on it.

In order to achieve this goal, appropriate data management and high-definition prognoses are essential: Where is the land? Where are the tenants? Where are they going to be? What services will they require? In our time, where access is more important than possession, should we continue to build as we have done, or has the moment arrived to start to think about flats like built-in community services? How will the demographic pyramid affect housing stock? How can we make all these questions more than rhetorical motives for study and turn them into operational questions that create policies? That's the key to Smart Cities, and to why the commitment to them is necessary.

A SMART CITY MUST DETECT FORGOTTEN CITIZENS

The fact that Patricia has to return to Ecuador is a collective failure. We need to confront those who applaud people like Patricia who leave as a consequence of our inability to offer them the future they dreamed of, with working solutions. We need Patricia and her hard-working, meritocratic ethics. We need her dream, and her dream to come true among us. People like Patricia are the new sap we need to avoid becoming a waning society, the worst face of a Europe that has buried its head in the sand. Only the city can offer safe ways to avoid giving up, to avoid the trap set by people that refuse to make theirs, what is truly ours: we're all participants in Patricia's frustrated dream.

A Smart city is fairer, because it has mechanisms to identify neglect, or the guile that others exhibit. Data also plays a crucial role here. Detect, detect, detect: that's Smart Cities' talent. When a city's operating system manages to cross-reference data and proposes solutions that effectively act on the problems, anticipating them and proposing, the city improves. It's amazing that people correctly receive packages from Amazon in their real houses but councils have a hard time even sending registered letters. The good news is that, thanks to technology and the advances in the construction sector, we know that cities can grow in a far more intelligent way than one hundred years ago. There is urbanistic consensus to determine what is the best volume to build and to determine the best densities. Here we have to go back to being truly open and fulminate prejudices. It's necessary to find physical models that maximise the quality of life in cities, and this will inevitably be done by constructing taller buildings or clustering them in a more compact way, just as the models developed using massive databases predict. Bjarke Ingels and his team have demonstrated that the best architecture in the world (see 8 House, Copenhagen, also known as the Big House) is possible, with the most powerful social purpose imaginable. Very important volumes embedded in a naturalistic, intergenerational and interrelational conception of shared space, where public and private are balanced in the civic. In the very same neighbourhood of Copenhagen, right next to 8 House, there are examples of some of the most dangerously boring social architecture.

A city is intelligence put at the service of collective management. Smart solutions, big data and following the pulse of the city in real time will transform public space accordingly. The design of the cross-section of the streets in Ensanche, done by Ildefons Cerdà, was a show of technology for the time. The façades of the sunny houses with wide pavements for pedestrians, tree-covered and separated from the road through which the most modern means of transport of the era would circulate: carriages pulled by steam engines. They are the same principles that drove Bjarke to maximise the light and pure air gathering properties of every apartment in 8 House.

Cerdà thought about everything: under the streets, large sewers through which the dirty water from neighbouring buildings, factories and facilities would flow. Large siphons would prevent the stench from rising to street level. Cerdà had spent a long time studying the state of Barcelona's districts (1840–50): a suffocating density of up to one thousand habitants per hectare; unsanitary streets without access to the sun; buildings without any kind of ventilation, residents who for centuries had been crammed inside medieval walls that were still fortified for military reasons: he had to stop Barcelona messing it up again. The consequences on daily life were terrible and incredible: sicknesses, epidemics, collective fatigue, social revolutions, ideological pendulums and flagrant inequality were just around the corner. Barcelona, the most revolutionary and most bombed city in history, admired by many, was awash with hardship, foul smells, misery and lacked space. Cerdà devoted himself to producing a detailed prognosis of the situation, gathering the big data of the time: mapping figures with pen and paper, as a few years before, in London, another of his predecessors had done, Dr. John Snow.

JOHN SNOW, CERDÀ, MOSES, JACOBS, GEHL . . .

It is worth reviewing John Snow's contribution to the science of cities. By the mid-nineteenth century, London's Soho district was suffering from a serious problem with filth, due to a large influx of people and a lack of adequate sanitation services. London's sewerage system did not reach Soho. Many cellars had black wells underneath the wooden boards which covered the floor. As these black wells were invading the city's subsoil, the government decided to dump waste into the Thames. This was the root cause of the outbreak of cholera that ravaged the city. On the 31st August 1854, after several outbreaks in other areas, a major outbreak of cholera reached Soho, considered the worst outbreak ever suffered by the city. Over the next three days, 127 people died in the Broad Street area. By the following week, three-quarters of the area's residents had fled. By the 10th of September, 500 people had died and the mortality rate had reached 12,8% in some parts of the city. By the end of the outbreak 616 people had died.

Snow was skeptical of the widely-held beliefs of the time. Common theories were that diseases like cholera or the plague were caused by contamination or a harmful form of "bad air" known as miasma. It's important to remember that Pasteur didn't propose the theory of bacteria until 1862, so John Snow was unaware of the mechanism by which the disease was transmitted. The evidence, however, led him to deduce that it was not propagated by victims breathing foul air. He laid the foundations of the scientific approach to urban transformation. In 1849 he published his first theory, an essay titled *On the Mode of Communication of Cholera*. In 1855 he published a second edition that contained much more elaborate research on the effect of the supply of water to Soho in 1854's outbreak. What John Snow did was very simple but very thorough: he gathered data and more data, collating the areas in which people had been contaminated with possible elements of dissemination. That's how he ended up at the Broad Street fountain.

Snow's revolution was the mass use of data to arrive at firm conclusions. He used pinpointed cases of cholera on a map to demonstrate how they centred around the Broad Street fountain. He also used statistics to demonstrate the connection between the quality of water at various fountains and cases of cholera. Snow's diagrams gave other collateral results; for example, the workers at a nearby brewery had not been contaminated by the epidemic. The fermentation of barley during the beer making process killed bacteria and eradicated the cause of cholera. Despite the evidence, it was hard to convince London's city officials that Snow's theories were correct. They were repelled by the idea that a public fountain could be the source of orofecal contact. It was the Reverend Henry Whitehead, a former champion of the miasma theory, who was convinced by Snow's evidence. He definitively tipped the scales in Snow's favour and a precedent for epidemiology was established. It was linked to urban metabolism, to neighbours' habits and to the aggregated analysis of a city's data.

From the example of the Broad Street fountain we can learn the following: John Snow had not set out to analyse the city as an integral phenomenon. His concern was epidemics, which is no small thing. In contrast, Ildefons Cerdà's idea was to visualise the city as a perfect human habitat. He didn't only think about streets and buildings. In fact, he designed buildings based on their function, again showing himself as ahead of his time. Cerdà, in the same vein as Snow, thought about what flowed through the city's arteries, he thought about urban metabolism.

In the period 1845–47 Cerdà the engineer, together with M.L. Marchessaux, managed the works that brought drinking water to the city of Valencia. For Cerdà this experience would be a key point of reference in the construction of Barcelona's Eixample (Ensanche in Spanish). In 1861 he assigned Josep Fontseré a map of all the existing drinking water pipes. This architect was another man who lived by mantras: "Don't destroy to build, but conserve to adjust, and build to make bigger." He had participated along with Cerdà in establishing the *Topographical Plan of Barcelona* in 1855 and the *Ensanche Street Plan* in 1861, bases of modern cartography. In 1870 Eduard Fontseré was commissioned to formulate a project for canalising, channelling and servicing drinking water to the city and its wider area, within the framework of the Barcelona Water Company's recent constitution of 1868. Veins, arteries and energy . . . Cerdà and his era were as systematically intelligent and Smart as we are today, it's obvious. We have talked about water, but the same applies to the arrival of gas. The urban gas service in Barcelona was installed by the French engineer Charles Lebon in 1848, to illuminate the city. Lebon relied on Cerdà to bring his gas company's interventions to fruition.

In his *Theory of City Construction*, Cerdà explains that gas wasn't used in England until 1805, nor was it imported to France until 1815, nor did it become widespread in Paris until 1829. Cerdà participated in the enlargement of the public lighting and private gas consumption distribution network in the widening of Barcelona, and he did it together with the engineer from the company M.L. Marchessaux, whom we have already met. From his position as a councillor on the city council, he helped the Lebon Company turn an isotropic city into a utopian reality supported by the most basic urban services in every corner of the city.

Cerdà marked the way by integrating water provision, sanitation, the supply and distribution of gas and telegraph systems into his studies. These were innovative services that would soon change from being unattainable to indispensable. Doesn't this story sound familiar? Additionally he thought about the different sections necessary to integrate the railway into the city, underground (he planned for the railway to pass through the centre of some of the new city's widest roads), as well as other modes of transport above ground. Cerdà planned the city integrating all possible dimensions, making possible the coexistence of housing with transport networks and urban services. If in Barcelona we dare to say that Cerdà was the inventor of urbanism, it's because he was the first urbanist to incorporate networks of combined urban services on a wide scale in his urbanistic proposal and in the scale of his project. A sort of Internet of Things of the era.

In light of this tradition, it's difficult to understand the resistance to integrating in the city the unique advances that have brought us the digital revolution. Smart Cities are not an adjective in a city's urbanisation. Smart forms part of our lives, and it has to move to positively condition the way we build a city. A healthy city, a sustainable city, an intelligent city, a happy city, a fair city . . . The aforementioned are dimensions of city that can be worked on to achieve urban balance. However, adequately balancing these concepts requires dealing with complexity and and counteracting the inherent of doing "as usual". Smart Cities are not a collection of marketing products, as some purport, but an advanced methodology to intervene in the city. A methodology focussed on *social* development, in which all social parameters are *measurable*, to be able to monitor and quantify them. It's an *adaptable* city, open to change, intelligently flexible and prepared for any change. It's a methodology that is *reproducible* in any other urban anatomy, and whose evolution is *traceable*. In short, a Smart City is an attitude.

S = SOCIAL
M = MEASURABLE
A = ADAPTABLE
R = REPRODUCIBLE
T = TRACEABLE

Data, data and more data! Intuition is arrogance, and a democratically responsible and mature city cannot afford to progress only on the basis of delusion, intuition and stellar projects. The ambition to transform is crucial, but the execution of change must be based on sound projects. Nothing is genuinely social if it does not take into account society's real movements. Years ago you could design using the "broad brush theorem," which consisted of a group architects that decided on a map, using a broad-brush approach, where the urban roads and housing estates would be. But today we know that this belongs to enlightened despotism and utilitarianism taken to the nth degree, and it only creates frustration. Decisions have to be backed up, and this involves increased knowledge as a product of the increased handling of data. Technology is a powerful ally in decision-making for the city.

On a rainy day in New York where we got lost in the Strand bookshop we stumbled upon Robert Moses. Because a city is also its network of bookshops. Moses was, like Ildefons Cerdà, another king of the self-commission. He was once a young and promising technician from New York: progressive, executive and dynamic. When mayor La Guardia appointed him NYC Parks Commissioner, he placed the Triborough Bridge Authority under his control, as well as six metropolitan agencies responsible for all of the city's parks, roundabouts and freeways. For many years this New York super-bureaucrat imposed his city model for cars. He was convinced that it was possible to combine the freedom a car gives, symbol of North American middle-class prosperity, with an aggressive urban policy of obtaining space to create children's parks. "Cars and parks" could have been his mantra. Getting down to work, and without Google Earth, Moses was accustomed to flying over the city in a small plane looking for opportunities. He was known for his ability to find the right space to develop this or that park, as well as positioning the bridges that theoretically had to decongest the more than congested already existing tunnels and bridges between Manhattan, the Bronx, Brooklyn and Queens. He was blessed by the press and they praised his hyperactivity, cutting ribbons left, right and centre. Its surprising to read comments about Moses, as the only man able to reconnect that macro-city, especially after having inaugurated the Triborough Bridge. Something had bypassed the press: none of the mobility studies that had been carried out were solid enough to justify that bridge.

Soon after its inauguration it became clear that the bridge hadn't solved the traffic problems, in fact quite the opposite, they had got worse. Moses hadn't used reliable data, but his intuition. With the willpower and leadership abilities that characterised him, Moses got to work again and decided to build another bridge, in this case the Bronx–Whitestone, he refused to allow trains to run around it, and finished the works three months ahead of schedule. In fact, the Whitestone Bridge had to form part of his Belt Parkway system that ran around Brooklyn and Queens. To carry out the work Moses entrusted his design to the engineer Othmar Ammann who introduced important technical improvements. The problem the bridge posed was that it meant destroying part of the Malba neighbourhood in Queens. It was the prelude to other truncations: later came Sunset Park in Brooklyn, and particularly the Third Avenue shopping district that was cut in half by the construction of the Gowanus Expressway. Cars were imposed with special cruelty on neighbourhoods in Harlem during the construction of the West Side Highway, which in fact cost New York its last urban forest bordering the sea, the destruction of a neighbourhood and the disappearance of the last freshwater marshes. The great builder had become the great destroyer, even though no-one, nothing had stood in the way of his triumphal march . . . There was no reasoning.

By the fifties Moses had lost his touch and someone more intelligent came along, endowed with profound astuteness. In 1952 when Moses decided to lengthen 5th Avenue, cutting through the middle of Washington Square in the heart of Greenwich Village, he didn't anticipate that its residents wouldn't be as resigned to the idea as those in Harlem, or as self-sacrificing as those in Queens. The opposition was led by well-educated residents, skilled at using the era's most modern social media. Moses, hellbent on using his intuition instead of objective data, lost the battle. The day Washington Square was finally closed to traffic Moses snapped at one of his collaborators, "There is nobody against this – NOBODY, NOBODY, NOBODY but a bunch of . . . a bunch of MOTHERS!" He was wrong. It was a sign of the times, a return to community values, to the social values of a city that were planted as a seed. It was, above all, the power of well-thought-out discourse. Jane Jacobs dedicated herself to observing, quantifying and proposing changes to the declining city using well thought-out theories. The use of sidewalks, the urban economy, the life and death of cities are not delusions, but descriptions of complex realities systematically revealed and comprehensible to any resident willing to get to know his or her surroundings.

Moses won the media battle because, in truth, all one has to do is see the world's megalopolises, where we do business or lose ourselves on holiday, to confirm that Moses' ideas are still in vogue: Dubai, Shanghai, Shenzhen and so many new cities are abased by the furore of those scalextrics, fly-overs and ring roads in their centres. In Mumbai or New Delhi it's plain to see how huge masses of concrete threaten the lives of hundreds of thousands of people that live underneath those depersonalising monsters that sever collective life. In Barcelona, in the first decade of the 21st century, a fly-over for high-occupancy vehicles was allowed to be built over another already congested road. Its designers were determined to light it up at night, because they thought their work was pretty. Didn't they care that nobody was using it and that by building it all the traffic would end up on the Meridiana, one of the most congested roads in the city? Neither Moses nor those responsible for the high-occupancy fly-over in Barcelona, appreciate data. And what's worse, neither of the two, each in their own way, seemed to have thought about what is expected of a city. Jacobs did, and for this reason her book *The Death and Life of Great American Cities* is still obligatory reading today, more than fifty years after its publication. By the way, Bennet Cerf, the book's editor at Random House, sent a copy to Moses. He sent it back in an envelope with a note that said, "Sell this shit somewhere else". He was left without an argument. Moses and anti-Smart Cities aren't so different: they lose value in their ignorance. They are essentially victims of their own fears.

Jacobs' battle with Moses continued until his dream of building the Lower Manhattan Expressway was shattered. It would have run along Broome Street and destroyed myriad historic, brick and wrought steel buildings, as well as a large part of Little Italy and China Town. Data reveals that the freeway wouldn't have added anything to the city, other than put more pressure on neighbourhoods already in precarious situations. Moses had forgotten the social function of his work, the human aspect, the need to produce spaces for creation and coexistence. Planning a city isn't drawing dots and arrows, as a plethora of intellectual architects and urbanists have conditioned us to believe. Planning a city is taking people into account and the millions of unexpected things they do every day. Here a couple of old men who weren't committed to anyone come into play, because they could afford to: a mayor who didn't seem like one, and an architect: Michael Bloomberg and Jan Gehl.

We met Bloomberg on one of those institutional missions to position Barcelona in the world and attract investments to the city. The mayor of New York received us in his office, a sort of engine room in the city where his desk was right in the centre, a huge room full of screens, deputy mayors, administrative staff and coffee machines. Bloomberg wanted to gather knowledge about the city and had put together an office that fitted his idea of what a mayor should be: someone close enough to the average person to know what they think and hope for in a political leader, and someone sufficiently informed about themselves and their team so that they know if the services they offer are in line with what was promised and the proposed vision. One of Bloomberg's revolutions was the mass use of big data and data mining to manage it. However, or perhaps as a result of this, when it came to making a city, to making it more intelligent, he turned to another unconventional classic, the old Danish architect Jans Gehl.

For a long time Gehl had been turning cities on their head. The data made it clear: people like living outside cities less and less. The Valley, in California, is emptying in favour of San Francisco. Students graduate from Stanford and move to the city centre. However big data also makes it clear that young people's spending habits have pushed cars into the background. Since 1990 the percentage of private cars driven by twenty-somethings has fallen from 20.8% to 13.7%. They are tired, but also waiting: tired of the smoke and noise, tired of time lost forever in interminable queues to get from one place to another, normally from home to work. But they are also waiting: waiting for genuinely competitive electric vehicles, at a good price, that allow us to drive without producing fumes, that allow us to move around in big shared buses, avoiding last centuries constructions, like building roads in the middle of a city. And then there's the value of experience: Gehl led the pedestrianisation of Copenhagen, in spite of its residents' and businesses' phobias and philias he managed to convince them with studies and data that a human scale development was possible and more comfortable in cities.

So, with all this data at their fingertips, as well as their intuition, Bloomberg and Gehl proposed transforming Broadway. How? Little by little, by subduing resistance to change, and by arming themselves with evidence that would overcome any reluctance. First they sent some interns to count people and cars to demonstrate that it was unacceptable to carry on living as they had been doing. I imagine Gehl talking with Bloomberg at that long table in New York City Hall, with two steaming cups of coffee, and telling him, "Alright, we're going to do it, but before we repeat previous mistakes let's do it right: let's leave intuition to one side and make everything black and white." They counted and observed, and they realised that Broadway's potential use as an intersection in Manhattan's grid system was striking. They closed Times Square and for the first time it deserved its name. They did the same with the corner of Broadway and 5th Avenue: they counted people and they counted cars. They closed a section of the street and they dreamed up a square in the densest centre in the world. Moses would be rolling in his grave!

"What? They've stopped letting cars in front of the Flatiron building? Those mothers again . . . " he would shout from wherever he was, without realising that if a mayor is capable of giving life to the city's epicentre, they will also be able to revitalise the suburbs that have been victims of deindustrialisation. Neighbourhoods full of victims of unemployment, of people from all over the world, of different types of families, of people with jobs that are never thought of, jobs that end up giving shape and meaning to things none of us have ever imagined.

There's a difference between proclaiming something and doing it: how many transformation projects have failed because of fear of change? An incremental transformation, in which the results are visible and the solutions adaptable as the space comes into use, is much more viable than starting from scratch. Gehl and Bloomberg's courage lies in taking the political and technical risk to change habits to improve quality of life, and doing so in a professional way. Their success in New York is a global manifesto in favour of human scale development and constitutes a benchmark for a new way of making cities.

MEASURE, MEASURE AND MEASURE, TO LATER INTERVENE . . . AND TO BE ABLE TO MEASURE SUCCESS

It's true: renovating streets runs a high political risk. Ex-mayor Jordi Hereu lost the elections due to the controversy and the fatigue created by the reforms of the Diagonal and it was difficult for ex-mayor Trias to reform without a majority. What's true is that now when people walk through the renovated section of Diagonal it seems ridiculous to have taken fifty years to give each method of transport its right to move: pedestrians have wide pavements, there's a service road and room for motorbikes, conserving the century-old trees. Now no-one remembers the hostility of the criticism against a reform that the Barcelona that doesn't like itself called "classist": the number of new businesses opened and new users of buses and bicycles speaks for itself. People now use Diagonal on foot more than by car, and the idea that it is one of the civic centres for excellence has been interiorised. We did it by putting up tables and hammocks, but also putting on concerts and entertainment for children on Sundays. We used civic organisations as invigorating forces, dedicated to social inclusion, turning the experience from opening up the city to a multidimensional action that strengthens social and communal aspects of life. We celebrated the conquest by collecting data on the appropriation of the new public space and we were pleased to see that the new distribution made all transport flow better, from cars to people. We did it by widening pavements, improving street-lighting, giving back to the other of Barcelona's Broadways, Paral·lel, to its theatres and its almost forgotten vitality, the urban walkway, with large squares and a green reservation for bicycles. All of these things sweetened the hard, port-bound avenue that the street had become.

Transforming Barcelona's Avinguda Diagonal cost 20 million euros, at a rate of €400,000 per year for the years during which the avenue remained unchanged. That figure is what some of the businesses located on Diagonal turn over every month, some every day. The current government team wants to establish a tramline, and controversy has surrounded the avenue again; the cost is calculated to be 175 million euros. With or without the tramline, there's no turning back from the change: Diagonal has become a civic core, its pavements can be walked up and down and it can be ridden safely by bicycle. Oriol Martori, from Urbaning, carried out the study that indicated the road's capacity for pedestrians and bicycles after the reform and all indicators of quality, comfort and safety multiplied their values.

Diagonal should serve as an example to broach other reforms that seem "impossible". Gran Via and Meridiana are the other two central axes that determine the city's morphology. A truly transgressive policy would be to approach changing these roads in a simultaneous way, with a comprehensive vision of what still remain urban highways. Diagonal should serve as an example to convince nonbelievers, instead of stoking the political fire between parties. Collecting data on the use of these roads as connectors on a metropolitan level should help to propose solutions for newly generated traffic.

Data on the use of public space is therefore a key application of Smart Cities and their transformation. So far we've talked about an extractive big data model. That is to say, we've talked about the city's datamine and started to look for the best and most attractive betas. We've put all the data into large silos, we've stored it in what have come to be called RDBMS or, "Relational Database Management Systems." This type of analysis was already common in the 1980s, even though tools for storing, analysing and calculating were far less sophisticated than they are today. Using these tools we've been able to better understand the Rambla de Barcelona. In fact, in great detail. We know how many people have walked down it, and when. We worked out what would happen if we moved a bus stop, or if we closed one side of the road off to traffic.

Knowing how to use public space is something that had obsessed us since Urban Habitat and has also worried district teams. For this reason, when Mónica Novoa, from *Foment de Ciutat* [City Development], asked us to help her understand mobility in La Rambla within the framework of the *Pla Cor*, literally "Centre Plan," we suggested she should work with Urbaning. This company, lead by a young multidisciplinary team, is a pioneer in the adoption of technology to monitor the use of public space. Through the installation of cameras that process their own images, and don't require the continual presence of operators, we were able to obtain its capacity: the volume of migration, composition of traffic, the direction in which people move and their hourly distribution. Obviously the treatment of recorded images complies with the rules set out in current regulations regarding data protection.

The use of Smart technology allowed us, through the placement of sixty cameras and the manual control of three strategic points, to gather information on the influx of people and bicycles to La Rambla, data that hadn't existed until then. The conclusions are very interesting. In the first place, we learned that 280,000 people and vehicles enter La Rambla on a working day. This is equivalent to saying that Terrassa and Rubí (on the outskirts of Barcelona), or four times the population of Palo Alto (California), fit inside La Rambla on any giving working day. At weekends there is a higher influx of pedestrians, 350,000. This explains the pressure that is felt on La Rambla: the high-density that is lived and breathed daily, above all by people on foot. The data shed light on shocking realities: more than 90% of the people that go to La Rambla do so on foot, only 6% take the metro and 1.5% go by bus. La Rambla functions in an unbalanced way: almost 60% of movement takes place in a longitudinal sense, with more people passing through the Raval (23%) than through the Gothic Quarter (18%). At the weekend these figures swing even more in the Raval's favour (27%).

The analysis, which has a ridiculous economic cost due to the value that it provides, analyses every crossing, all migration out of La Rambla, and the hourly distribution of influxes to the zone. But what's really important is that it detects "hot" zones, where the greatest discomfort is felt, where pedestrians are forced to slow down and change direction. Physical contact is inevitable and changing direction almost impossible. At weekends, at peak times, there are several locations along La Rambla where it becomes impossible to walk and convene in comfort. This is a visionary method of public space analysis because it compresses analysis times. Of course, with a team of one hundred people and a few months out on the street these behaviour patterns could be detected. However, on the one hand, the presence of "people watching" would inevitably misrepresent spontaneous behaviour, and on the other, it would be incredibly expensive in terms of time and cost.

The study was a turning point in the La Rambla project because the data allowed the architects to understand the problems, to identify them on their plans and to consider interventions. La Rambla has an identity that has been well established by time and does not need a redesign that might further harm it. What it needs are better crossings, to move urban furniture so that it doesn't interfere with the flow of people, to minimise the number and types of lampposts and bollards, and to get rid of its unbearable pavements at its thinnest points. All this should be done by making better use of space (the volume of pedestrians decreases the further you get from the Plaça Catalunya and closer you get to the sea), and enhancing the residential dynamics of adjacent neighbourhoods. For example, we discovered that it would be a mistake to pedestrianise La Rambla entirely, because the carriageway on the Raval side is a fundamental access road for the internal distribution of cars in the neighbourhood, whereas on the Gothic Quarter side it is used as more of a thoroughfare.

If the best design for public space is one that is functional, that serves people, that aligns the flow of people with the places they want to go in a comfortable way, then this type of analysis is fundamental for the new method of planning public space. Let's hope that no-one decides to implement a solution by radically changing the layout of the road, because with little investment a few junctions could be remodelled without losing the identity of one of Barcelona's most characteristic features. The 300,000 people that pass through La Rambla daily deserve an excellent design, focussed on its purpose and its livability. Let's make sure that this is La Rambla that goes down in history.

The use of big data has allowed us to work on the surroundings of Barcelona's Sagrada Familia as well, and to understand the extent to which the intensive and invasive use of its streets has suppressed the area's residents' daily use of public space. Having a tourist attraction in the heart of Barcelona like the Sagrada Familia is a blessing . . . and a massive problem for those who live close by it. In the seventies my brother and I used to skate in one of the parks that are to the side of the building, in the one that's next to the Passion façade. My grandma and aunt lived very close to the basilica. Neither of the two dreamed of seeing it finished, and in fact they doubted if we would ever see it finished because the basilica can only be advanced with contributions from the people who visit it, theoretically to expiate blame and sin. At that time there were even a few people who wanted to tear the basilica down. Today, however, it's difficult to find a focal point as important as the Sagrada Familia.

In the 21st century, the use of big data 1.0 allows us to have a large, analysed and structured database to be able to locate sources of congestion, traffic jams and talk authoritatively with the residents affected by this permanent invasion. You don't need to live near the Sagrada Familia to take a certain dislike to it. Anyone who walks past it, runs into the streams of organised tourists or sees the line of coaches will be reluctant to go back. Or for anyone looking for a flat in the area, rents have shot through the roof due to the pressure of tourism. Or anyone who takes notice of the ground floors of the area's buildings, whose restoration has been scandalously poor, or the number of souvenir shops that have spread like wildfire. But at the same time the Sagrada Familia is a site of massive importance in terms of cultural, architectural and religious heritage for the city, so we have to roll up our sleeves and refine possible solutions based on the evidence shown by data.

Off the back of these experiences we proposed systematising our methodology and widening the reach of our data collection. Big data 2.0, produced by platforms like Google, receives feedback from the mobile devices we all have in our pockets (3.0). 2.0 gave us the possibility of working with data from social networks, it allowed us to maintain a direct connection with residents and the civic world. At the same time, it allowed us to keep in-depth analysis of the city's heartbeat up to date. 3.0 gave us more volume. Let's not forget that with ten billion mobile phones expected to be in our pockets by the year 2020, and with more than 80% of the people that go to the Sagrada Familia using a smartphone, we will be able to monitor people's every step by tracking devices that connect to public Wi-Fi.

We still have a long way to go: what about the ground floors? What do we buy, what services do we need, what are the profiles of people that come to live here, and of those who leave? The data, big and small, is very valuable given that it allows us to correlate statistics about requests for public services, the habits and needs of citizens as a whole, but also individually. Cross-referencing this data with sociodemographic data, and other types of information, allows us to predict better, to design better, to anticipate dysfunction before it occurs. Data explains the present and the future of a city, by employing it we are able to learn and act simultaneously.

One of today's most pressing debates is about the gentrification of the city. It can be seen in Barcelona, but also in London, Mumbai and New York. Public and private investment in a city tends to significantly increase its assets, and as such its prospects of capital gains. This effect attracts capital and incentivises social mobility. However, on many occasions this mobility entails the ostracism of the area's old inhabitants, ejected because they can't afford to carry on living in a place which only the rich can afford.

Barcelona is a good example of a city in which public investment has drastically changed certain neighbourhoods and districts. Perhaps the most obvious example is the invention of beaches as a public space, more specifically the beaches at Barceloneta. This had traditionally been a neighbourhood full of industrial workers centred around the company *Maquinista Terrestre i Marítima*,[4] of fishermen and of dock labourers. The closing of the company in the seventies, along with the fishing crisis and decline in port activity, plunged the neighbourhood into a deep economic depression made up of unemployment and hopelessness. I will address this problem later in the book. The area was plagued by a massive influx of drug use and crime which stigmatised one of the city's most characteristic districts. The Olympic Games provided the perfect excuse for heavy intervention: the city decided to recover (perhaps it would be better to say *invent*) the neighbourhood's beaches. It's plain and simple: behind the Barceloneta, or in front of it, depending on the angle you look from, Barcelona installed the world's best urban beaches. The direct consequence is well-known: after a few years, almost undetected, yet there for all to see, the tourism industry had taken over a large part of the neighbourhood's activity. It's not exaggerating to say that everyone was taken by surprise. What?

4 A Barcelona based metalworking company, founded in 1855.

The fisherman's neighbourhood, with the beach bars, in the hands of international tourism? Well yes, international and national. The sudden agitation could have been better controlled if they had put Barcelona 3.0's resources at the service of the mixed model that some people defend.

We got to work, and through the necessary in-depth analysis of the available data, big and small, it didn't take long to discover that along with a let's say classic tourism industry, Barceloneta was becoming one of the most important world hubs of the maritime industry and the mending, restoration and winterising of large yachts. The yachts are what you see, the part that turns some people's stomachs because of the luxury they represent. What you don't see are the people that have daily work in Barceloneta thanks to this activity. Those that design boats, fix machinery and repair the port. There are 300 companies linked to the maritime sector in Barcelona that generate more than 800 million euros per annum and close to 8,300 jobs. Barceloneta has 16,000 residents.

Industry is the alternative to sun and beach tourism, the type of industry that these workers make tick, directly or indirectly, dedicating themselves to a key sector of a city that lies by the sea and has the power to activate its coastline. The majority of these companies are small and mid-sized, employing an average of fourteen people. The challenge is to make both worlds communicate – the residential Barceloneta and the industrial Barceloneta. Both sides have an enormous responsibility to make sure that the neighbourhood genuinely participates in the development of the Clúster Nàutic.

The Port Vell marina is located in Barcelona's port, between the Ronda Litoral and the passeig de Joan de Borbó and is the fruit of reform started in 1989. It's a public marina and a yachting port. The marina is more than 25 years old and has old facilities that are very expensive to maintain. This motivated the company that had held the concession since 1992 to sell up to another international group in 2010 for 30 million euros. This is the price that the PSC-ICV[5] coalition put on this public space until 2021. In spite of the demagoguery, we don't think it's a bad deal and defend the Clúster

[5] Partit dels Socialistes de Catalunya [The Socialists' Party of Catalonia] and Iniciativa per Catalunya Verds [Initiative for Catalonia Greens].

Nàutic as a valid alternative to touristic monoculture. Marina 92 has the most sophisticated spraying hangar in the world. Here, ships of all sizes are painted in a submerged hangar. This is the part of the port that is not widely known, and that also employs people from Consorci El Far. This is a public organisation created in 1993 that promotes all forms of culture to do with the sea. It develops training programmes, and educative and outreach programmes that are linked to the nautical and maritime sector. We would never have committed to the construction of the iconic W hotel on Barcelona's public coast, it changes the panoramic views of the beaches and streets of Barcelona and brings more pressure from tourism to the area (there are many other points along the metropolitan coast where one or more buildings of this type would play the dynamic role that we urgently need). Given that it's already been built, we applaud its commitment to employing people from Barceloneta. You can negotiate in the city, of course, but *with* it. The maritime industry is much more than thirty luxury yachts. However to combat prejudices you have to collect data, systematise knowledge, and share it.

The data gave us clues as to how to intervene, how to keep attracting investment, how to consolidate what we already have and how to propel it towards the future. With the data we started to work on the intricacies of the neighbourhood, giving incentives to responsible entrepreneurs. In short, the data allowed us to take action with policies that favoured a productive city (getting back in touch with the mantra), that favoured the creation of opportunity, fighting against uncontrolled gentrification and allowing an endogenous variety of actions that afforded local families the opportunity to better their living conditions at the same time as seeing their assets, what they own, what they can bequeath to their families, increase in value.

When it comes to planning a city, institutions are a good thing until they're the only source of transformation. Turning Barcelona's coastline into beaches, forgetting the industrial sector, was the result of institutions from the Olympic generations. However a more detailed look at the uses of regenerated neighbourhoods would have allowed the prior detection of mass tourism's dangers. In the same way, a more detailed look at the uses of public space would allow us to design more fit-for-purpose areas, the spaces that belong to everyone, shunning prejudice and subjective observation. When Moses anticipated the revolution of the automobile he anticipated the right to free movement, but his vision was unable to achieve complexity or encompass finer detail. Moses wouldn't have liked to see what his models, based on freeways and bridges full of cars, would have produced thirty years later. We arrived in time to intervene in Barceloneta, but if someone had done it before us we would have gained time.

6

Innovation, Manufacturing Associations and More Social Big Data

INNOVATION: AN INTEGRATED PROCESS

During our years in government, Barcelona was named the first European capital of innovation. The nomination's really something. We can't simply ignore it as if it didn't exist. At a time when voices around the world were heard talking about the need to reorient ourselves after the world economic crisis, and whilst we were emerging from it, the European Union recognised that in Barcelona we opted for change. Redirecting the focus of our productive, intellectual and social capacities was key in creating new jobs, and attracting and retaining talent. Although Barcelona, almost always unsure of itself, didn't give this idea its due importance.

Perhaps the reason is that our commitment to innovation could have easily become a *commodity* policy. Innovation can turn into one of those empty concepts that filter into language from time to time without anyone really knowing what they mean, but they sound good. We're tired of hearing politicians from across the world, here in Barcelona as well, who call for equality, equity, rights . . . innovation, when they don't really know what to offer. You might say, it's better to talk about innovation than gambling or bachelor parties, and it's true. But talking about innovation for the sake of it is like saying nothing, and it ends up devaluing the concept.

Innovation entails solving complex problems that have no practical solution, and fostering new economic approaches for cities. Using these two ideas as focal points we created a strategy that we developed on many fronts (businesses, academia, civil societies and institutions) and it turned us into a global benchmark. Furthermore, whatever happens, whoever directs the city's destiny, the road is still open and there's no way back. We managed to create a joint leadership focussed on a simple yet very ambitious vision: to use technology to better understand complex problems, fine-tuning appraisals and encouraging all sectors to collaborate, proposing ground-breaking solutions.

Academics from all over the world, prestigious authorities, large and small businesses, experts, lecturers and scholars state far and wide that during these years Barcelona has been one of the most open cities to innovation, risk and challenges. We have to carry on doing it. International businesses made a firm commitment to Barcelona, like financing the Smart City Campus in the old Ca l'Alier[6] factory, in the most neglected part of Poblenou. For once, the Barcelona miracle, in which a mix of quality of life, a wealth of culture, a culture of wealth, the understanding of leisure's potential, working with pleasure and the city's innate vitality, produced a unique combination of intellectual challenge and a vision for the future. The Smart City Campus was affected by everyday doubts. Meanwhile, Berlin has taken the initiative: Innoz's great Smart campus project, lead by Florien Lennert, has taken the lead by attracting investors and city research centres.

To comprehend the innovative potential of a Smart City, lets go to the Col·legitel Nen Déu. It's a school run by nuns, dedicated to children with psychosomatic, psychomotor and autistic disabilities. We went there once with director of the Mobile World Congress (MWC), and we did so to show him what our commitment to innovation implicated. Sister Roser received us, a wonderful woman, with quantum-like powers (she's capable of being in two places at once doing completely different things), logically she was wearing her habit for work. We explained to the director that it had been Sister Roser and her team who had shown us the path to Social Smart Cities, a path that we had spent a long time looking for. Sister Roser showed us through the school's classrooms for children with different types of handicaps: some were canteens, some were for lessons, others for gym. . . . In all of them they felt the need for electronic tablets that would make the lives of so

---

[6] A previously disused factory that produced textiles in the Sant Martí district of Barcelona.

many people with some kind of disability a lot easier (after all, who doesn't have one?).

Connected tablets are these people's voices. A tablet is a chair's remote control. A tablet is email, a way of communicating with transport, with siblings, with parents and grandparents. A tablet is the tool that lets them say they love us. And we really need them to tell us! Tablets supply information on health and mobility sensors. This information applies to each individual student and to all of them at the same time. Individual interaction and immediate aggregation. That's the key to what Smart Cities are!

We told the international director that the real MWC was in the Col·legitel Nen Déu, and schools and centres like Esclat, or like the Escola Jeroni de Moragues and so many other centres and occupational workshops in which the technology associated with an intelligent city and mobility make Barcelona a better city. With this philosophy Barcelona won Bloomberg Philanthropies' Mayors Challenge, and many other awards that helped entrepreneurs to develop transgressive projects for citizens with real problems.

However it wasn't only this. Barcelona managed to certify its innovation management process. Innovation isn't an isolated concept: it's a process that starts with creativity and ends with achieving results. When we received the UNE166002 certificate it was a great milestone that allowed us to measure results and above all control the innovative process; just because you're innovative doesn't mean you cannot be strict. This exactitude was the cornerstone of achieving the title European Capital of Innovation 2014–16.

MANUFACTURING ASSOCIATIONS: THE GREAT TOOL FOR THE DIFFUSION OF DIGITAL CULTURE

The digitalisation of urban life is changing the way we use the city's public facilities. The world's first Ateneu de Fabricació [Digital Manufacturing Civic Center] was inaugurated by more than one thousand families that registered on the internet. Since then, the network of Ateneus hasn't stopped growing. We owe it to two civil servants: Pep Garcia Puga and Jordi Reynés. Don't assume we're talking about young people with dreadlocks or technology focussed studies; we're talking about fifty-somethings, mature men, perhaps slightly stuck in the 20th century, but people with a great desire for the 21st century, for innovating, for relearning how to serve people from public service. They worked hand in hand with the FAAC.[7] They invited families from Barcelona to experience the pleasure of digital manufacturing of making robots and working in different programming languages. All of this might sound very strange, or perhaps you might think that the families that went to the Ateneu were special. That wasn't the case, these were families that were in the Ateneu on Saturday, and at the Fundació Miró or the Teatre Regina on Sunday, or simply going for a walk or a run down the Carretera de les Aigües. These families, however, were curious to learn the language that will mark the 21st century, the century of hyper-connectivity and the Internet of Things. The path, as with many other things, was opened by the Institut d'Arquitectura Avançada de Catalunya that I founded in 2002 with Vicente Guallart, Willy Muller and Manuel Gausa. Together we discovered the potential for the digitalisation of the city, and the transformation of daily life.

To sum up, it's been widely agreed that the invention of the steam engine transformed a society based around agriculture (economy 1.0) into an industrial society (economy 2.0). The changes were so profound that today there are still parts of our society that haven't completely digested them. Both speeds coexist, although development has accompanied the latter far more. The widespread access to education that accompanied the capacity for the generation of wealth, the ethos of reconstruction and redistribution in the postwar world, as well as the subsequent increase in standard of living and expectations of social comfort, instigated the birth of the service economy (economy 3.0): education, health, leisure, commercial services started to became real economic driving forces.

Even today we can still find people who talk about health or education as if they were expenses, when investing in health and education is investing in essential generators of wealth. What's clear is that the evolution of the economy, the cycles from which society regenerates itself, are becoming shorter and shorter. The *.com* economy, based on the internet, on *www* and mass social media (economy 4.0), turned

[7] Fábrica Automatismos Apertura Cancelas [Gate Opening Automated Systems Plant]. Producer of cutting-edge automation systems.

distribution channels on their head and opened the door to a different relationship between products and consumers. Consumers' freedom of choice grew exponentially, whilst manufacturers' distributive and business capabilities had to be completely rethought. In short, we saw how the Amazons, Googles, Wikipedias and Open Universities changed how we bought, sought information and learned.

We're making the last leap forward without realising it. It's a combination of the phone we put in our pockets and the digital decoding of everything we own, do, and consume. However it's also the way in which we communicate and make decisions. Until recently, the dream of downloading digital audiovisual content from the internet seemed like a miracle cure. Nowadays it's normal. If I were to say that relatively soon we'll be able to download 3D content . . . What would you think? Well, "We cannot predict the future but we can create it," as Abraham Lincoln said. With computers that will be as intelligent as the human mind within ten years, with the unstoppable big data wave, with the shift in paradigm created by digital printing or decentralising production of the majority of what we consume, and therefore changing the concept of personalisation, society will reach a point of no return, the digital revolution, the economy 5.0.

This new era will scrutinise the previous four from top to bottom. It will question them and integrate them. Resolving problems is intrinsic to technology. The cities that best identify challenges will be the ones that take the most advantage of technology. For example, a city could locally produce the majority of what it consumes: from clothes to food. It wouldn't just be urban allotments, it would be creating green façades on buildings, renovated rooftop terraces as productive community spaces, producing electricity or tomatoes. As well as intelligent restoration and green spaces. Underground spaces could be used to store energy in batteries and release it at the pace of demand. Industry would have to adapt to the relationship between production and consumption; the internet and phones have given us a new social grammar, that of decentralisation and the personal configuration of the definition of product characteristics. Services are already offered in an individualised way, and .com channels incorporated into phones have left for dead those that used www as a launch platform.

Much has been written about the danger of commercialising Smart Cities. What isn't in danger of being commercialised? But subscribing to a technological service is a personal choice. It's true, having an app on your phone with the Ikea or Mango catalogues can make you consume more and more clothes. But, I insist, the choice is personal. The same choice exists for second-hand products, with applications like Wallapop, the very popular platform that allows you to sell second-hand objects through a simple process that doesn't take longer than thirty seconds. It involves taking a photo, describing the product and putting a price on it. Whether from your sofa, bed, or on the bus the app shows you available products, organised by distance from you, price and chronology. In the same way we can request a book from a public library or swap books with neighbours.

Those who tend to be against Smart Cities feel threatened by its business model for new initiatives. The Luddites, English boatmen who tried to sink steam-powered ships, are rearing their head again. The opposition to evolution always has its roots in fear and mistrust. Manufacturers of buggies, for example, are worried about their competition: it's incredibly easy to buy a second-hand buggy at the touch of a button for less than half the price, without the middleman. The government is suspicious because the rules of purchasing have been changed, they've threatened to raise taxes. If it's not about a challenge to the "as usual" way of doing things, how can the hostile critique of platforms like Wallapop be understood?

The digital revolution is shaped like a new world. Cities can aspire to be leaders in the implementation of this revolution 5.0 which will start to make the old world redundant. This revolution has the potential to create thousands of jobs. Everything's still to be done, to be produced, to be exchanged, to be recycled, to be transformed . . . to avoid the city getting left behind! This revolution consolidates freedom based on the democratisation of knowledge and it marvels us every day with its capacity for collaboration that allows us to live more cohesively and communally. The thousand families that participated not long ago in a day for children at the Ateneu de Fabricació de les Corts aren't an anecdote. They're the present turned into hope. It's the 5.0 city that's able to reinvent itself without fear. Knowledge is the basis of trust.

Knowledge and trust go hand in hand in the Digital Manufacturing Associations. The Manufacturing Associations are creative and educative spaces that encourage the use of new technologies where citizens, organisations, universities and businesses can get involved in a digital creation laboratory that allows the community to make their ideas and concepts into a reality for useful projects for society. They have multi-functional facilities, are open-plan, made of reused architecture and large in size. They're equipped with precision cutting machinery, 3D printers, computers, tools and circuits and they operate in a network, there's one in every district.

The Manufacturing Associations' objective is the following: to break barriers, to get people closer to digital technology, and to help them understand and become comfortable with it. The Associations are a collection of buildings that share resources and teaching, but in a few generations we will have all this equipment at home, we will have integrated digital manufacturing into our routines and we will be able to programme from home without having to depend on a third party. They encourage social innovation and a new model for industrialisation. They allow citizens to create products, ideas and solutions through the stimulation of an open and shared knowledge in which the people who stimulate are the protagonists. In this way, the Associations' users are co-producers of the innovations that will mark the city's development, promoting a culture of collaboration orientated towards a change in the production model.

The Associations can also provide career opportunities for students who aren't able to work within the framework of regulated education but feel attracted by the power of seeing their creations produced by 21st century machinery. In many cases, the ideas aren't theirs: knowledge exists and it is shared on the internet. However, they digest it, they absorb it, they assess it and they produce it. Once the product is made, they evaluate it, examine it, give it some thought and improve it, sharing the result with the network.

There are technology parks and universities all over the place equipped with similar machinery to what you can find in an Ateneu de Fabricació. In Barcelona there are many private co-working and digital creation initiatives. Leaders in social innovations, like Cecilia Tham the founder of Mob, have inspired the creation of the public network of Ateneus. As well as the pioneering FabLab from the IAAC (Institute for Advanced Architecture of Catalonia) in Poble Nou, or the most recent one in Can Valldaura, in Cerdanyola. What's disruptive about Barcelona's model is that these facilities, whether public, private or a mix, collaborate with each other. The Manufacturing Associations don't stimulate a change in the production model; they create the community that is going to. That's why our Smart City is a social dream of collective dignity. It was using this dream, that transcends generations and party politics, that we made Barcelona the world capital of the new way of thinking about a city, definitively centred on people, through technology. A simple but truthful revolution that creates opportunities and jobs for Catalonia's capital, for the European capital of Innovation.

Barcelona's commitment to Smart Cities hasn't been without its critics. Some time ago someone wrote that the problem with the commitment to Smart Cities was its emptiness, its lack of content. These people argue that technological solutions applied to the management of cities "create more problems than solutions." When someone who is theoretically well-informed voices an opinion of this type it becomes, without meaning to, a symptom. In this case it's a symptom of prejudice. It's true, Smart Cities can be a commitment lacking in content if innovation policies' objectives are not well defined. Another symptom is talking about social cohesion, without doing anything effective to achieve it. Prejudice is easy, quite obvious and very common: for years people have been saying that computers are dreadful machines, that they take up a lot of time and that life was better when we wrote everything by hand. The reality is not only that we have grown accustomed to computers, but that we now carry them around in our pockets. The technology sector has had to get on the right side of its users and it has made our lives easier, because they haven't had any choice. Education and training for all generations (not only the younger ones) on how to get the most out of technology has also helped its expansion.

The digital revolution first occurred on a domestic scale and then was translated to the city. First the internet was introduced on desktop computers, in schools, universities and businesses, in government and finally on the street. The future of cities is going to be more Smart because their inhabitants already are. From the kitchen to studios, shopping to reading, the internet has become the first source of information people consult.

Velocidadcuchara.com, for example, is a webpage written by a cook who shares recipes from the internet, anecdotes, and secrets for lovers of sophisticated cooking . . . with a Thermomix. She has thousands of followers that write to her, compliment her recipes from years before and keep up to date with her patients in hospital (she's a nurse). The blog is a simple, well-designed wordpress with the photos and videos that she takes. Nurses, cooks, singles, and elderly people send Rosa comments daily, congratulating her on the dishes that she prepares with her food processor or showing interest in different ways to make the recipe. Like most successful blogs, it's an interactive platform and the author draws from the requests, comments and information that her followers give her to choose the next menu or dish. She creates "challenges", shares recipes with other blogs and interacts with her fans and providers, perhaps even charging for it, using a business model through which she shares content to promote the best of other brands.

Can you cook without blogs and a Thermomix? Obviously, but it's clear the people that make the recipes find the webpage useful, stimulating and creative. It's also free, and might even be a substitute for their old cookbooks. Her free blog is far more successful than the recipe books, platforms and official applications created by the brand of food processor she uses; her platform is named after the machine's speedy mixing.

When technology is applied to a city the same sort of thing happens. The majority of *velocidadcuchara.com*'s users haven't got a clue about what a Smart City is, nor would they be prepared to pay for new technology-based services, although they use apps, smartphones and councils' webpages everyday to keep up to date with what's going on in their city. Furthermore, if doing so makes them more aware of the city, gets them closer to the council and doesn't cost them anything, they'll be radically in favour of using technology to provide more efficient urban services and improve their quality of life. Well, this is why a Smart City has no competitors, because it's a paradigm shift that goes beyond some misguided product of technology companies' marketing, it gives each citizen the value that they decide to invest in it. In addition, the citizens themselves aren't only passive users of information that the administration feeds them, but they are the principal content generators.

Let's be fair to those who fear the change created by Smart Cities: the problem lies more on our side, the side of those who are on technology's side to turn it into human development. Perhaps we haven't been able to explain *why* properly. We've made it too easy for rhetoric. We've forgotten the value of facts and we've let ourselves be swept away by hot air.

The time-honoured Ramon Llull has already correctly taught us that although faith may not be able to be easily changed, understanding can be. In other words, it needs to be understood that in the world of the physical and communal construction of our cities, there's a massive space for humanity to grow *through* the development of digital technologies.

BIG DATA AT THE SERVICE OF A PRODUCTIVE CITY

Through, by way of . . . We consciously choose the words we use. Smart Cities are not an end in themselves but a means to an end. This is their great value. Through their development we have to propose joint objectives with direct consequences on our lives. For example in Barcelona, once the Olympic model which had benefitted the city so much between 1992 and 2004 had been used for the thousandth time, we decided to take a step up in scale. We distanced ourselves from a model focussed on investment in large infrastructure and growth through mega-events to a model that applied technology to improving the intelligence of the city that was already built. It was always clear to us that the commitment was a way of overcoming the devastating effects of the financial and housing crises from 2007–2011. However at the same time, it is a way of dealing with the crisis affecting current city models that are based on a passive one-dimensional understanding of cities. This typically 20th century vision did not take into account that the digitalisation of reality allows for analysis on a new scale, and therefore totally new prognoses and action.

Our commitment to a productive city, that's human-powered, hyper-connected, energy self-sufficient and with zero emissions needs to have a very strong technological base, founded on data and a great ability to innovate. Why? To create a new economic-productive sector to give citizens better answers. That's the other great value of Smart Cities: they generate new economy, they open society to a new productive ability.

Solutions from the 19th and 20th centuries have proved ineffective at addressing the challenges of rapid urbanisation and inequality. Sustainable urbanisation will only be achieved if we can surmount the current model of development. Improving life in cities is an objective that requires sharing knowledge and administering a good dose of innovation. In order to humanise cities again, the use of technology is essential, it's been a force of change in cities from the beginning: from

Greek temples to skyscrapers and the latest social housing programmes, development has always been possible thanks to advances in construction. But it is not enough anymore. Paradoxically we need more technology to combat the very defects that development has created. We move around the city more, we consume whenever we want and we depend increasingly on the rest of the world. We need more data and more intelligence to make development compatible with urban growth, and traditional tools are not good enough.

To consolidate Smart Cities as tools that break the mould, the private sector has the responsibility to understand collective challenges and orient their solutions towards citizens' real needs. When dealing with complex problems, data provides objectivity: sustainable development and a fair city involve universal connectivity, open data, and a digital recruitment strategy. Public services should serve people, and not the other way round. For the first time Smart Cities incorporate prognosis, the ability to take action, personal interaction and social action. Smart Cities afford citizens a more active urban life. The urban landscape will change as a consequence.

In 2015 Catalonia had 3,400 businesses in the ICT (information and communications technology) sector, which is equivalent to a quarter of all ICT based companies in Spain. Almost 70% of the businesses in Catalonia's ICT sector are located in the province of Barcelona. The high capacity, security and capillarity of the available ICT infrastructure in Barcelona and its metropolitan area position the Catalan capital ahead of other Mediterranean countries. Almost all Catalan homes have a mobile phone (90%), computer (73%) and broadband internet (67%) at their disposal. This reality had enabled Barcelona to give a distinctive value to Smart due to its integral approach to the concept. The Smart concept is normally employed as a link to sustainability: cities like Copenhagen have fused the descriptive "Smart" to the adjective "green," and they've championed tools that normalise the city's ecological footprint.

In Barcelona we tried to apply the "Smart" approach to all areas of the city, making the best use of technology to tackle complex problems using a cross-field method. In fact, all the city's projects were developed using joint leadership from various departments. Housing, Infrastructure, Environment and Urbanism worked in coordination with the ICT Department to solve problems that had been looked at from a one-dimensional point of view until then.

Barcelona has an advantage because it has a 500 km-long fibre-optic telecommunications network that has more than 700 public Wi-Fi connection points. This integrates the city's different networks whilst increasing capacity and capillarity and keeping costs down. Furthermore, with conveniently timed legislation (we forced ICT integration into urban transformation), we obtained a network of sensors distributed throughout the city that provide continuous information about the its rhythm: movement, transit, capacity levels, traffic lights, environmental conditions, noise, hotel occupancy, the spread of tourists, etc.

Innovation and technology applied to the city and to the services that it offers us are synonyms for hope. However, this is a perception that has been filtered through the lens of Google Maps, from top to bottom, a lens that makes it difficult to see people's faces. Sometimes it's comfortable to view things this way, given that big figures disguise what really happens in people's daily lives. It's similar to being on a plane and studying the landscape from high up: we know what's down there, way down, there's people like us. We see an organised, postcard-like landscape, and we ignore the pain and avoidable suffering that it conceals.

If we look through a Street View lens we find: firstly the technology associated with social care, for example for elderly people living alone. We also find tele-health for people with mobility difficulties, technology for people with any kind of disability, technology to improve the quality and efficiency of public transport, and technology for waste collection and treatment that improves our ecosystem and as a result, our health. Let's not forget the technology that helps us to better understand who we are, or that allows us to better plan the city and even build and maintain in a way that costs far less than it would have done a few years previously. Not to mention the digital education distributed at all levels of society and what it means for our children, our adults and our elderly. We can't forget, as this book demonstrates, that the introduction of technology 2.0 in our daily lives allows us to add and remove data at will. On the one hand this implicates having instant analysis available of the situation that has been added, and at the same time it allows a situation that might only affect one person to be managed. This ground-breaking flexibility, as is so often said these days, or integrative and natural flexibility, however we prefer to qualify it, converts the Smart City phenomenon into an opportunity. An opportunity that we need to contemplate in the same way that people contemplated access to running water in their homes one hundred and fifty years ago, or even one hundred years ago, when the Mancomunitat de Catalunya proposed installing telephones across the country.

Smart Cities may have not made us any more intelligent, especially some people, like myself... What's true is that in Barcelona we realised how to lead the investigation and develop new services and advanced diagnostics that would improve people's quality of life and create new opportunities – but the result is still unknown. We attracted talent through human resources and job creation, we pulled in productive international investment, we became the innovation capital of Europe, we created more companies and start-ups in the technology sector of cities whose solutions already serve their citizens, we saved energy, we fixed CO_2 emission rates, we saved water and we elevated the perceived quality of the city's services to higher levels. In addition we accelerated the transformation of the city in favour of human scale and economic activity. This is the true value of Smart Cities.

SOME SMART FORMULAS FOR MOVING FORWARD

There have been pilots, show-cases, city labs and all kinds of discursive devices put at the service of different projects, all of them legitimate but for the most part far from having clear public intentions. What hadn't existed until now was an authentic public strategy, that was holistic and integrated to open the city up to technological intelligence. At this stage of the crisis there are some lessons to be shared. The commitment has been a resounding success. Only with real learning, internalising the lessons and evaluation of the impact of the policies we launch, will we be able to face the future with guarantees. Here are two or three that might be useful:

ONE: *Only countries with economies based on knowledge move forward. This isn't new. Catalonia is a country that has almost only relied on its people's intelligence to move forward.*

When we talk about knowledge we're talking about technological advances, cutting-edge design, renovated productive processes, fresh governance. Ultimately we're talking about economies that haven't gone to sleep, waiting for the good times that don't necessarily have to return. We need to be up to standard across the board to be up to standard economically; this means progressing with the state of arts, science, technology, law, design and all disciplines, whatever they may be. Previously economies struggled against obsolescence by closing themselves off or devaluing the currency. Do the formulas sound familiar? None of these are viable today.

The digitalisation of production methods facilitates a direct relationship between us and them, changing the paradigm from the individual to the collective: it can be seen in the *uberisation* of services, with the loss of the middle-man. Above all, the potential lies in the ability to solve problems in a collective way, directly from the individual to the solution.

The crisis is systemic. The President of the Republic of the Congo said so, so did the previous President of the United States. Obama aimed high and launched a national initiative by which, by law, following Barcelona's example, all the states had to have a Manufacturing Association. We're not talking about a crisis that can be solved by sticking a few plasters on it. If the crisis is systemic then systemic solutions are what's needed. This means that we have to get rid of the parts of our economy that don't add value to the whole. We have to do it without the trauma of the savage restructuring that took place in the eighties. We have to do it with a scalpel. We have to do it while caring for and looking after the people that will suffer as a consequence of the change. To do so, the human scale management as provided by Smart Cities is fundamental.

TWO: *It is necessary to open new paths in the field of energy, making inroads towards energy self-sufficiency and zero emissions.*

What would become of Europe if, from now on, we proposed turning it into the world leader in solar power, for example, or in hydroelectric power? A dream? Didn't Finland dream of becoming a global benchmark in ICT? The economic system's breakdown forces us to take strategic decisions that have a journey that will last at least 20 years. Only in this way will it be possible to get out of this current dead-end. I'll say it firmly: pollution isn't a whim. Nor is the fight for energy sources.

Energy is a source of well-being. We buy tonnes of petrol because it provides us with tonnes of well-being. We, westerners, have used and abused this method. We cannot come to the fore, so we simply say that the world is going down the drain. What are we saying to the farmers of Andra Pradesh that have just got electricity for the first time? The only option is making the alternative better than the current way of achieving well-being through energy. That's the only way we will end the suffering that we cause the world, and that we cause ourselves – importing energy and exporting instability. The only solution is to move towards clean energy self-sufficiency. Starting with cities.

THREE: *We have to create a new communal agreement, a new social agreement based on the digital reconversion of a large part of essential methods of production, starting with a reconceptualisation of the collaboration between people and organisations.*

The agreements reached in the 1970s have become exhausted. The western world has entered a new phase after the golden years, after the nineties and the first decade of the 21st century. Extreme individualism, social insensibility, the gregariousness of the state, and the large private and public organisations' permanent custody over the individual, should give way to a new agreement that harnesses the best qualities of individual ability, which we've now learned is incredibly powerful, and the best of group strength, where everyone shares responsibility and profit.

The middle classes and their taxes have rescued banks and financed debt across the world. It should be with ambition and courage that, emerging from the crisis, we create a new contract with these productive classes, and with the most disadvantaged. A contract that allows us to save costs on innovation, share advanced tools and complex software, as well as institute fairer mechanisms of financing services and infrastructure that don't commit us to pharaonic investments – all toward the end of substantially improving life in cities. This is the *sine qua non* condition that will guarantee democratic quality in our societies. Smart Cities will also help to guarantee this condition. We can see it in Barcelona, but we can also see it in India, or Colombia, above all in cities like Medellín.

MEDELLÍN: THE FIGHT FOR THE CITY

You have to read and reread *The True History of the Conquest of New Spain*, by Bernal Díaz del Castillo (the first great modern journalistic chronicle), to understand the extent to which the past haunts Latin America. The *encomenderos* [owners of indigenous peoples] left a trail of despotism, insurrection, and as Aníbal Gaviria the ex-mayor of Medellín once told me, a trail full of the spoils of war. Medellín is a sunny city located in a wide and spacious valley, but it gives off the aura of being a secret, as if the plateau and mountains that surround it wanted to hide it from the world. This could be one of the reasons why, since colonial times, Medellín has been a city associated with alternate realities: the untamed tropics that cover the landscape with intense green; the search for El Dorado whether by miners, coca growers or those responsible for the great mafia clans that were prolific in the nineties. Medellín's geographic isolation, nicknamed Colombia's silver cup, seemed to destine it to marginalisation, despotism, and inequality which are the cornerstones of any totalitarian regime. In fact, Medellín, sunk inside a sort of green crater, lush with palms, ceibas, passion fruit and all kinds of tropical flowers, has the air of a lost paradise, an amnesia-filled bottleneck, and therefore, of a dream factory. Who knows if this is the reason why the conquistadors, the miners looking for gold, the coca growers, the guerrillas or the cartels have been obsessed by a cosmic obsession that has turned them into life's corsairs.

At present, a strong and noble generation of Antioqueños,[8] led by politicians with vision, are waking up from a harsh reality made up of affliction and tiredness of despair. Through Medellín's veins, its ravines, these indifferences and the memories lost to amnesia, descend. Through them flows the culture of looting that is also displayed by the big corporations that raffle off small human communities . . . But Medellín decided to instigate change and opted for a new urbanism radically orientated towards social development. It invested in the future, committed to Smart Cities, to different urban transport, to data and quality of life. In Medellín the self-built neighbourhoods are hyper-connected to the rest of the city through the Metrocable[9] and

[8] People from the province Antioquia, to which Medellín belongs.

[9] A system of cablecars that reach some of the informal settlements on the hills surrounding the city.

through a powerful network of libraries equipped with books and technology that bring culture and digitalisation to any corner of the city.

Medellín has become the example of how leaders' sociopolitical willpower can instigate powerful changes. There are few cities in the world that have been capable of coming to so many consensuses in such a short amount of time. In 1991, 7,500 people died a violent death in Medellín. Today the figure has been reduced to one thousand – still painful, but full of hope. Medellín was Escobar's cartel's city. Pablo Escobar put a price on the heads of politicians, journalists, and socially responsible businesses. Escobar bought goodwill and indulged in cheap populism whilst he put bombs in Fernando Botero's sculptures, who denounced him using the language of the people through deformed bodies and expressions frozen in time, just like the corrupt morals of the traffickers.

Today Medellín is remaking itself through innovation and prestige. Barcelona's alliance to Medellín was very strong. The feeling was mutual. From the urbanised city, to the urbanising city, and vice versa. That's how the world is currently divided. The truly meaningful division. There's an urbanised world, in which the implementation of all kinds of policies starts with the advantage of having a strong base from which they can have a positive impact in a reasonable amount of time. In the urbanising world, the suburbanised world which we find in neighbourhoods of Abuja, the shanty-towns of Mumbai, the ghettos of Philadelphia, the suburban ant's nests of China and the French *banlieues*, life is much more difficult. But we can start from the premise that we all have the right to hope. Intervention in the communities, the immense shanty-towns on the slopes Medellín's valleys, sucked out poverty with the installation of the metrocables (the cable cars associated with metro network), facilities like libraries, schools, public and privately funded medical centres, children's parks . . . but also by doing business, as is the case of the women of Morro Moravia. From a mound of rubbish they cultivated a magnificent crop of flowers which they sell to the whole city. Of course there are some people that criticise the fact it has become a business, because business is a source of inequality, they say. The protesters are a few men (amongst them there don't tend to be any women), who wait with open hands for someone to buy their support with a few breadcrumbs.

The challenge that the two cities proposed between us was making our shared vision of a socially intelligent city, a Smart City applied to social progress, become a transformative reality, in both the physical city and socioeconomic life. Medellín and Barcelona are today the two world capitals of urban innovation thanks to the work carried out for a few years, in large part jointly, with mutual respect and admiration. The vital part of this transformation was to combine to scales: long-term vision with actions that facilitate change. The challenge faced by cities like Medellín and Barcelona is resolving the contradictions between citizens' aspirations (to have more comfort, quality of life and economic progress) and the need to avoid endangering territory. Both cities want to improve their level of well-being without creating inequality. This can't be done without knowing who we are. Without having the necessary data.

WHAT DO WE KNOW ABOUT OUR CITIES?

Barcelona, a pioneer in the world of intelligent management of complexity, has the challenge of using this technology on causes which matter. The great intellectual provocateur that is Miguelo Betancor presses us to act with the type of questions he poses from Las Palmas de Gran Canaria to the world: Where are the most vulnerable people, and how can we get closer to them to improve their daily lives? How many empty flats are there really in the city and what methods would be effective at making them available for people that don't have a home? How many elderly people live alone and rarely leave their houses, and how can we develop a culture of healthy exercise and social relationships? Where are indebted companies concentrated, and how can we help them improve their situation? How can we help to create jobs in the formal sector taking people out of informal sectors? How can we remove barriers to encourage the use of public transport? How much food do we throw away everyday in the city and how can we encourage a widespread culture of quality nourishment and healthy life? And how do we do all this relying on the talent of the digital generations, shirking the analogue prejudices of the more mature. They're very relevant issues for which the collection of trustworthy data is essential, because the first thing that requires a consensus is a correct prognosis. The correct use of this data adds value to the city, saves resources and allows us to dedicate ourselves to what really matters. This adventure that Gran Canaria is starting towards becoming the first intelligent and digital island in the world has become meaningful due to these questions.

Big data is essential for predicting behaviour or anticipating conflict. In 1997, the archaeological site of Pompeii was declared a World Heritage Site by UNESCO, as a "complete and living testimony to the daily life of a society in a precise moment of the past." The eruption of Vesuvius in the year 79 BC froze, to this day, a compendium of testimonies of the era's daily life that have allowed archeologists to precisely decode the city's activity on the day of the explosion. Today it's possible, with expert help, to identify brothels, electoral propaganda or to learn the Roman's most sophisticated construction techniques. It's a fascinating task for archaeologists to reconstruct life from two thousand years ago. It raised the following consideration: why don't we dedicate the same effort to knowing how we live in cities today? Understanding the complexity of urban life is of anthropological interest and above all practical, it should raise a lot of public interest. Imagine a street in Barcelona, Carrer de Muntaner. What do we know about Carrer de Muntaner? And Carrer de Picalquers? We have a lot of static data at our disposal: length, width, number of trees, we can count the windows of all the buildings, we could calculate the volume of paint we would need to paint all the façades . . . In contrast, we know very little about what really matters, which is the people that live there, how they live, what awaits them. Administrations have devoted a long time to managing cultural heritage and the built environment, because they're the only things that could be achieved. Data has been historically hard to obtain: from tax collection to the census, following the population has cost the government lots of resources. If archeologists have been able to reconstruct images of mummified Pompeiian bodies and obtain a high-resolution photo of the moment the volcano froze time, then today it's possible to obtain thousands of high-resolution photographs of every minute in the city.

We register data with every step we take. We upload it to Facebook. We spread it via Twitter. We publish it on LinkedIn to form part of the working world. By combining data, we can obtain high-resolution images of the city. We collectively expect the government to go beyond its regulatory, judiciary and reactive roles. No-one wants elderly people to live in their city without contact with their family or friends. It's also intolerable that children skip school, and it's not fair that there are informal activities that are invisible to the tax system or people working without social welfare. How do we want to change the future if we don't know the present well enough? Finding answers to issues like the rampant unemployment rates that have crept up on us requires knowledge, a lot of knowledge. And it requires collaboration between sectors to decipher dynamics that are incomprehensible from a single perspective.

Theoretically, governments don't know anything about what goes on inside buildings. On the inside flats are for their owners and they're spaces beyond municipal concern. Its true, according to Roman law, private property is not the responsibility of the administration. But in the 21st century, what happens inside private property and what happens to the people that live in a city has an effect on it. The data's there: there are thousands of flats photographed from every possible angle to advertise their rent or sale online. If the administration doesn't have a department for "The Insides of Housing," it's because it doesn't want to. The tools are there, the data is free and 100% trustworthy. Just with the information that a site like *Idealista.com* provides, users can know if the property has been renovated recently, how many bedrooms it has, how many have windows it has that look onto the street, which rooms get direct sunlight during the day or if the kitchen is gas or electric. The owners don't skimp on any detail: they explain if the roof terrace is communal or not, if they think the building is quiet or not or if the flat is furnished when renting. If, in addition to all the photos of flats, the main entrance to buildings were required to have a floor plan on them, we'd have a more precise map of the city than we have today. We would be able to find out the average size of bedrooms, if people prefer them facing the street or the interior of the building, we would know how many bedrooms in the city don't have ventilation access to the street or to the building's courtyard, and we would be able to come to conclusions about people's eating habits according to their kitchen surfaces and number of people in each flat.

Data on the main entrances to tourist buildings is used to sanction the owners of apartments for tourists without licences; but the data has constructive potential far beyond sanctions. For me, we're not making enough of this value. Barcelona Council launched a website to report the presence of suspected illegal tourist apartments in buildings. It was meant to be the zenith of tipping off the council, but it didn't work. They got it wrong. It would have been ground-breaking to work with individual providers and owners to determine their action in the city and its implication on the development of the streets; it would have been ground-breaking to encourage tourists to buy from local businesses, and to visit the side of Barcelona that is sometimes invisible. To make visitors aware that the Sarrià, Porta, Can Peguera and Horta districts are as interesting as Ciutat Vella.

People aren't stupid, and apart from the hooligans, no-one likes to spend their holiday submerged in an ocean of shouting, noise, lack of peace, and activity all day long. Neither do the tourists that come to visit us and live with us. That's why the city of San Cugat is working in a pioneering way to develop a platform that combines street lighting, digital identity on mobile phones and leisure and rest friendly policies, with Smart technology. Using digital tools to publicise the metropolitan area, make clear its connections with other parts of the city and encourage a more polycentric style of tourism would be a more intelligent, Smarter policy than the fines, sanctions and visceral campaigns against tourism in general. Knowing more will allow us to fine-tune better, evaluate correctly and optimise resources to allocate them where they are most needed. This is the idea behind social big data. Behind Smart Cities. That's the great value of detailed number. High-resolution cities are the great value of the Smart social commitment.

7

How to Organise the Smart City?

HAL IS ALWAYS AROUND THE CORNER

2001: A Space Odyssey is still fascinating. The film was shot almost fifty years ago, and it sums up some of the seemingly dwindling obsessions of the sixties generation. Above all, the ineffable transcendence of human beings and their fight for knowledge and truth, whatever it may be, just as Pontius Pilate asked Jesus of Nazareth. Much has been said about the film, especially about the meaning of that rectangular monolith, a sort of conceptual Moai from Easter Island placed on a movie set. Its perfection, the smoothness of its material, the reflection of light, its sudden presence in situations apparently unrelated to its meaning, like when it appears amongst *Homo sapiens'* predecessors, have turned it into an ineffable icon . . . However, as the years have gone by, along with more screenings of the film, the film's centre of gravity has shifted to HAL. Who is HAL?

Let's go back to the final part of the film: in the year 2001, the spaceship Discovery is travelling towards Jupiter with five members of crew. Three in hibernation and two awake, David Bowman and Frank Poole. They are accompanied by a heuristically programmed state-of-the-art supercomputer with "eyes" (whose lenses are fish-eyed with IR sensing capability) and "ears" all over the ship, called HAL 9000. HAL governs the ship using artificial intelligence that allows him to talk to humans through speech. Life on board is monotonous: the two crew members that are not hibernating eat, sleep, exercise a bit, play chess with HAL, regularly communicate about how the mission is unfolding, have trivial conversations with their acquaintances on Earth (with a seven-minute delay due to the distance) and little else. Nothing is said about the mission, apart from that it's an exploration of Jupiter's environment. Just before arriving at their destination HAL 9000 asks David Bowman if he has any doubts about the mission and its secrecy, but when he responds negatively (instead of replying with a shade of doubt), HAL interrupts himself and announces a failure in a communications unit that could mean losing connection with Earth. David Bowman leaves the Discovery to go for a space walk in one of the small spherical ships with mechanical arms, to replace the theoretically damaged unit (a parabolic antenna that synchronously keeps the spaceship in contact with communications centres on Earth). When he gets back inside, all the analysis done on the unit comes back as positive, and mission control, a twin computer of HAL, also doesn't find any fault, which puts HAL's reliability into question.

Bowman and Poole take precautions so that HAL doesn't hear them talking about the possibility of disconnecting his superior functions, but HAL manages to understand the astronauts by reading their lips. HAL re-diagnoses the future failure of the replaced unit, taking the decision to reinstall the original unit to check whether it is broken or not, but at the same time to confirm his reliability. This time it's Poole that makes the substitution, but when he leaves the small spherical ship, HAL takes control of it and throws it at the astronaut, killing him by depressurising his spacesuit and hurling him far from the Discovery. Bowman isn't sure what has happened: HAL pretends not to know and Frank doesn't respond. Bowman gets inside another of the small modules, or small spaceships with arms, and goes to rescue his colleague. With Bowman outside the main ship, HAL kills the hibernating crew. Bowman manages to reach Poole, but he's already dead, so he takes him in the arms of the space module under his control and goes back to the Discovery. However, HAL doesn't open the hangar, he stops him getting in and warns him that without his spacesuit's helmet the trip could be deadly.

He tries a risky manoeuvre that HAL doesn't think he can pull off: he leaves Frank Poole in space, and with his small ship's articulated arms opens the exterior emergency door of the Discovery (whose manual opening is outside HAL's control) and he points his ship's hatch towards it. He then activates the hatch's emergency self-eject function and is launched inside the Discovery, where he is able to pressurise the access chamber. Now with a helmet, to avoid the depressurisation caused by HAL, Bowman goes to disconnect him. On route HAL seems to want to explain what has happened, apologising for everything and asking not to be disconnected. He is able to enter the room that contains HAL's principal functions and David starts to extract memory cards one by one. Little by little HAL's functions are disconnected. HAL begs David for compassion but with more and more communication problems he seems to return to his primordial functions and first teachings, like in the song *Daisy Daisy* by Harry Dacre (1892), and finally he is deactivated. When HAL is disconnected, a pre-recorded video of Dr Floyd starts to play and he informs David of his real mission: to investigate the origin of the signal received by TMA-1 in 1999, emitted from a point near one of Jupiter's moons.

In an interview with Stanley Kubrick published in *Play boy* magazine in 1968, they asked him about the possible interpretations of his film, *2001: A Space Odyssey*. Kubrick's answer was clear and at the same time encouraging enough for an army of hermeneutists to engage in interpretive research over the years, "You're free to speculate as you wish about the philosophical and allegorical meaning of the film . . . but I don't want to spell out a verbal road map for *2001* that every viewer will feel obligated to pursue or else fear he's missed the point." We are the gullible people that have tried time and time again to figure out the meaning of the monolith: transcendence, extraterrestrials, the energy of the universe, God . . . But, I was previously saying, after watching the film several times I have come to the conclusion that HAL 9000 is the film's real protagonist – that stubborn mainframe who knows everything, because he thinks everything depends on him, even the lives of his astronauts.

HAL 9000 is the best metaphor for bureaucracy and hyper-functionalised organisations. This applies to the private world of large corporations, as well as the public world of large administrative apparatus. On occasion public entities have been privatised from within the public system, although it might seem untrue, on behalf of a few civil servants that end up believing they are the chosen ones to handle the city's administration, as well as the country's and the democratic institutions that lie on the peripheries of the people's will expressed through elections. HAL is always around the corner, above all to forbid creativity or change. He does so through self-serving logic. He does so through enormous intelligence, through exceptional knowledge of his surroundings and his functions (those that have been created for and by him). But also through an infinite distrust of anything new, anything uncontrollable; in short, anything that for us represents a city open to change.

The strategy used by HAL 9000-style organisations comprises dividing themselves up as much as possible, to create hyper-regulation based on ensuring the traceability and strength of administrative acts. In addition there is always a lawyer in the back ready to sow seeds of doubt about those who dare to innovate: the organisation is the organisation, and everything that escapes it could be potentially dangerous. Someone explained this to me plainly: the administration is called that because it is designed to administrate. If it were oriented towards inventing, it would be called the innovation; or if it were at least organised to implement projects it would be called project implementation. As simple as that. Therefore when a political team has a real desire to transform, it's imperative to act very quickly without giving concessions to HAL 9000 or political influencers of any sort (something we can always end up becoming!): to understand the objectives and available forces, look for allies, and become legitimate in the eyes of the numerous civil servants that are truly looking to free themselves from the ruthless yoke of the HALs that can be found all over the world.

It's possible to isolate HAL if you have a clear vision, an action plan and the desire to put pressure on the system, to transform beyond the mere act of administrating. The science of cities is inexact, the reason why it is impossible to find two identical municipal organisations even if the genome of the city integrates the same components.

ORGANISE YOURSELF ACCORDING OBJECTIVES. BE ADAPTABLE. BE STRONG

The first thing that's important is understanding the complexity of the machinery, understanding its components. We summarise into a generic blueprint, applicable to the DNA of any city, and from this, outline the organisation necessary to launch a good Smart City project.

Apart from their personalities and living conditions, which differ and define them, all cities share the same elements: a natural environment, the same critical verticals: water, energy, waste, mobility, nature, ICT. And all of these verticals group, interact, interfere and occur in public and private domains, in the flats or houses we live in,

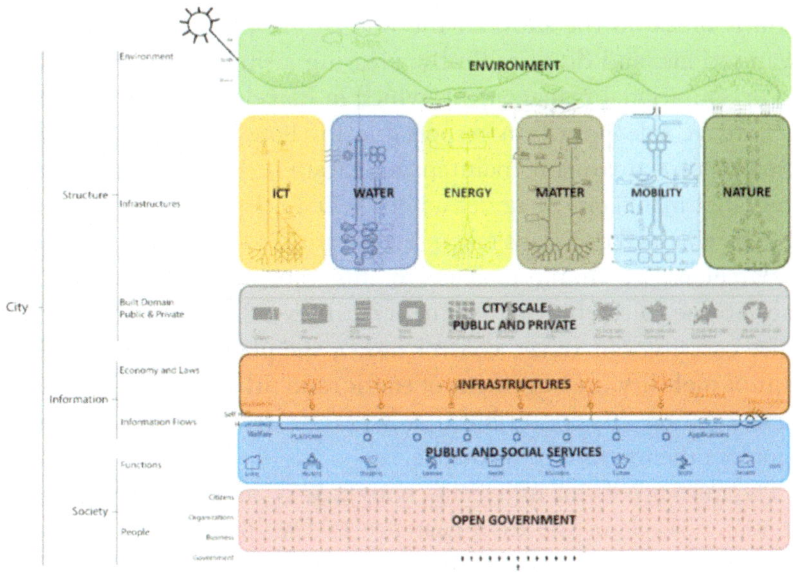

they condition our quality of life and the quality of space we all share: our squares, streets, pavements and all sorts of places in which we recognise ourselves as citizens. Without infrastructure and public services we can't talk about cities either. What happens is that, above all, necessary services are able to reach our homes in a more productive and agile way than ever before and give us personalised help, designed for every one of us, for every one of our needs.

As we are seeing, what's truly Smart doesn't take place on a keyboard, or on a phone. The Smart revolution happens in people, which is where we have to focus our organisations, organisations that are so introspective, so focussed on being HAL 9000. Hence, on the back of what we're doing with eGovernment, that was talked about so much during the first decade of this century, we are not taking advantage of the social opportunity that technology gives us. If a digital government only serves to technologically maintain and reinforce the closure and privatisation of public entities by a non-elected class, which a few high-ranking civil servants may well reach, then we're more dead than Frank Poole. Digital isn't a synonym of intelligent, the latter is provided by human capital.

The system itself is eager for change. Every election, every change in government means uncertainty and change, but predominantly the hope that things will work better. Therefore the first thing you have to do is break all the silos, all the water-tight compartments of our organisations, and do so radically, creating a single unit of political and organisational management, which in Barcelona we called Urban Habitat. A sole politician responsible for housing, mobility, ICT, infrastructure, urbanism, maintenance, cleaning, energy, water, parks; integrated management under an efficient leadership team; a shared table at which all decisions taken revolve around the mantra . . . Well, in Barcelona we didn't manage to integrate mobility into Urban Habitat due to the blurred liminality between discipline (municipal police carrying out traffic control tasks), the planning and management of mobility (management of traffic) and urban maintenance and transformation. It's not serious if, as was the case, dialogue and shared vision are the marks of governance.

The makeup of this unit isn't enough. It is imperative to organise the team based on the disappearance of the intrinsic dynamics of departments that normally work separately, perhaps even competing between themselves due to personal distaste. In Barcelona we started to discover some of these incommunicative spaces that made rolling out strategies for change or urban transformation at the same pace as other times extremely difficult. In fact, a part of the city's Olympic nostalgia resided in the fact that in that era, the end of the eighties and the start of the nineties, the vision was followed by ad hoc organisation that made changes happen incredibly quickly. Vision, budget, organisation, all aligned with well-oiled collateral resources.

Nowhere is it said that those who work in the public sector are susceptible to flexibility or adaptation. It's true that many years of doing the same job (whether public or private) ends up stultifying anyone. It's about shaping groups whose trademark is their malleability: their malleability and their adaptability. And at the same time consistency. I would almost say strength . . . provided it's full of that much needed and very rare malleability. Everything's a lot easier when the leadership conducts itself with transgressive willpower, that is to say, they give themselves legitimacy through their team's consistency in order to start making reality evolve. This demands intelligibility, transparency and above all, making the team feel part of the project. Forming part of a project is one thing, but collaborating to make a project possible is an entirely different matter. Municipal organisations shouldn't neuter the unique quality that is provided by a mix of politics (ideas, vision, compromise, passion) and functionality or public service (consistency, traceability, technical ability, knowledge, security).

COMMIT TO INNOVATION, FORGET THE ORGANIGRAM AND HARVEST TALENT WHEREVER YOU FIND IT

I've spent too long talking about something that is too similar to an organigram. My intention is by no means to confuse: Organigrams are far removed from our concept of modern organisation. We could almost say that organigrams, in their current form, are the chief enemies of our way of understanding organisation. An organigram is an excuse, a pretext for those who don't know how to lead. If you have a load of boxes, a load of levels, arrows that go from one side to the other, you've definitely got a good hiding place at hand. And that's fine if what you're looking for is to hide. Organigrams have become the greatest comrades of the modern HAL 9000s. They hide public positions that are specialised in privatising jobs and the public budget. They're those people that treat elected politicians as caretakers, as the British series *Yes Minister* showed us so well. Fortunately they are a minority, but we all know these officials, theoretical civic servants who are prematurely honoured with a little sign on their door that confirms their level and job, and they cling to it until they've privatised everything: they always have technical arguments available for paralysing other people's projects. It goes on everywhere, in hospitals and universities and in the multinationals as well, where our alma maters become infected; however, indolence, lack of vision and selfishness are particularly serious in municipal structures. Organigrams are the enemies of creativity and innovation. In fact, one of the greatest tricks that our beloved HAL 9000s play on us is convincing us that they're creating an innovation department: a director, a few deputy directors, a nice section in the corner of the mother of all organigrams, and in four days into the innovation director forgets to innovate . . . because what's most important is keeping that heavenly administrative position of comfort that the administration has give him, behind closed doors.

Innovation, creativity, imagination are isotropically scattered across the administration, and we know it. Careful, because HAL knows this before you do, and he'll fight to stop creative innovation taking hold of the administration. Because . . . it's difficult to promote innovation! It's difficult to conclude that the fusion of Urban Habitat is reflected in a lamppost or a housing complex concurrent with opening up the city for works. It's difficult to accept that all housing policies cannot exist without the aggressive acquisition of public land beforehand, making deals with private landowners to reverse the market's inertia whose only interest is that everything carries on, "as usual." It's difficult to realise that the best showcases of the World Mobile Congress or Smart City Expo and World Congress will never be in the Feria de Barcelona, however important it may be, but in a school for disabled children. All of these examples of innovation, that are real, and that we have reviewed in this book, led Barcelona to become the leading European capital in innovation, appointed by the European Commission.

The innovation race goes back a long way and forms part of all institutions, even the most traditional ones. Religious orders, for example, have been a source of inspiration for many organisations. Bishops that have been very important in our recent history have graduated from Solsona.[10] For example, Bishop Deig or Cardinal Tarancón, a key figure in the Spanish Transition to Democracy, who managed to stand up to the fascists of Franco's dictatorial regime and worked to achieve democracy in Spain. Tarancón, just like Deig, combined theological sense with popular sense, and for this Solsona is a good school. An afternoon in the sun in the plaza de Sant Joan of Solsona is testimony to this. Deig was a good footballer, and a massive Barcelona fan. For him life was about teaming up. His episcopal motto was, "*fer, fer fer, deixar fer*" [work, promote work, let others work]. A good leader must know how to do things on their own, how to make their team work, that is to say to incite action, and they have to instil creativity, that is to say, to let their people act, let them grow, let them enjoy the improvement of their jobs and the products and services that they offer. In the administration's case, this isn't obvious.

Without limiting ourselves to Deig, our cultural roots are grounded in learning history. In a world heading towards self-help, a hyper-secular world in which anything that seems important is subjected to summary sentencing, there are lessons that we tend to

---

[10] A town in the Province of Lleida famous for its cathedral and episcopal palace.

forget. From age-old organisations such as the Benedictines or the Society of Jesus, we can deduce various lessons, useful for applying intelligence to improve the organisation of the great mechanism that is the city. From Ignacio de Loyola[11] and his followers, we've learnt that the most important thing is a person's knowledge of themselves, that is to say, knowing what rules to abide by when thinking about your followers. The good ones, but above all the bad. In the same way, to value the genius of those around us, trying to integrate it into each and every one of our work processes, making sure to impose ingenuity on the "business as usual" way of thinking that ends up curtailing any possibility of progress in the administration. Subsequently, love and humour; love for work and the people we work with. Love for what we do, yes, but especially for those who do it, which means rewarding them and giving them a helping hand when they need it. And to let us help when we need to. After so many years of working on the front line of politics we have kept many friends, the majority of whom have never voted, and will never vote, for the party we work for.

We start with the fact that working for the city is, or should be, a shared motivation for all municipal employees, and that this excitement shouldn't be dampened by the complexity or fatigue that people who work on the front line are subjected to by the media and political parties. Good urban leaders should know how to concentrate on what is important, what motivates staff to love their job, beyond the day-to-day life of the leader that can seem lonely. Judgement, the renunciation of resentment, the ability to unite two seemingly opposed ideas, even if one of them is not a reflection of our political or ideological sphere, help us to get the best out of ourselves and our democratic institutions. Work ethic must form an integral part of leadership.

Good organisations are based on the conformation of elites that share an explicit and well-known objective. I say elite not in the sense of stock or lineage, but as a synonym for competitiveness and meritocracy. It's the vision and aim of those who propose changing a city, a country. Tony Judt, the great British historian, accused Western

[11] A Basque Priest (1491–1556) who founded the religious order Society of Jesus in 1540.

society of having confused the idea of equality of opportunity with the triumph of mediocrity. A team that leads must have the courage to propose ideas, without having to please everyone. Debates in which everyone agrees don't bring about any meaningful conclusions. Mediocrity is comfortable, but it's an enemy of change.

Furthermore, organisations are only as good as their members, hence it's imperative to ensure that people are incorporated and remain in them through methods of continuous assessment. Giving people easy entrance and life-long tenure weakens the whole organisation. Merit and experience are fundamental in any organisation, and even more so in those that serve society. In recent times we've seen how some platforms have had a complete disregard for experience of any kind. It's like getting rid of the knowledge we have accumulated for generations by recording mistakes. The most innovative projects require the experience of those who have spent many years working to overcome resistance to change.

Organisations should be as slim-line as possible, with as few levels and channels of communication as possible. Hierarchisation, hyper-formality and the cult of the leader destroy the conciliatory and liberating capacity of an organisation. This is often at odds with short-term interests and senseless partisanship. We must be consistent until the end: this paradigm, these very pages I am writing now, are going to be controversial very soon, even for those who might agree with them today. On the other hand, hopefully senior figures will have the sufficient critical capacity to innovate organisations. And for those who are on the outside, to change their paradigms that seem to be set in stone.

An organisation's ethics cannot be forced. Ethics have to be connatural to the organisation, which moreover has to equip itself with the necessary means to immediately reveal the practices that could corrode it from the inside out, without too many people realising. Without personal ethics, legitimisation will not be possible.

It's imperative to take decisions about responsibilities by basing them on merit and not on appearances or influences. It's essential to take them away from the people that constantly complain, those who are never happy and are always tired. It's important to seek the support of people who are jovial and enthusiastic, those who speak to the mayor and his subordinates in the same way. It's so obvious that I almost don't need to say it: what's important is giving lessons on the back of what has been done, not what has been said. Credibility is the first thing that's lost. All kinds of miracles are promised during the campaign, but the figures will give you away from the first year of government. You can be extroverted, talkative, humorous, serious, committed, earnest, humble . . . all at the same time, without sacrificing one characteristic for the other. In any case, you have to be a good leader, and that means listening before speaking, understanding without being obliged to share. A leader's capabilities lie in the creation of collective ambition, not in his own.

The key lesson, the one which overcomes HAL, the one that allows us to move forward at a great pace, was surrounding myself with direct collaborators and organisations committed to the common good, with good governance and management oriented towards increasing the quality of life across the world. All of this whilst fighting against sectarianism, governing for the people without wanting to be liked by everyone, uniting, building bridges, getting everyone on board, putting up with rubbish and admitting mistakes. In Barcelona mayors have the power, and this is quite exceptional on a global scale. The power of Barcelona's mayor has been gained by setting an example, through the strong leaders that have honoured the position. Today it's only possible to govern in this way, where you stand with one foot among top brass, planning, and the other in the trenches, in the line of fire.

However, the mayors who call the shots must know that they cannot do so without founding things, or without relying on ad hoc agencies that are independent of political vicissitudes. We relied on a public limited company, Barcelona de Infraestructuras Municipales (BIMSA), to whom we gave the authority to work, work and work. We also reinforced the Instituto Municipal de Informática (IMI) and its operational capacity, making it the centre of operations for the development of innovative projects and smart projects. We could say the same about a company like Barcelona de Servicios Municipales (BSM), responsible for a large part of the operational deployment of Smart strategies relating to mobility, parking and logistics in general. Businesses and agencies governed exclusively by professional criteria and focussed on services. Reinforcing these aspects was what earned us praise in terms of quality of services and citizens' perception that quality of life was improving. Independent professionals that are separated from the day-to-day running of things, for the sake of day-to-day activity, and for the sake of strategic projects. With regard to this, Barcelona is undoubtedly the model to follow.

From this point on it's important to learn, question and educate. To invest in excellent training for civil servants, to encourage continuous learning in a wide range of subjects, to arouse interest in what happens in homologous cities abroad, to learn about other realities, to confront dogmas, to become aware of what we know and what's left to know. A council employee that doesn't know about other public policies or projects in other cities is like a doctor that only has one patient: they end up conniving themselves that the diseases are unique and that there's only one good way to practice medicine. There are no good and bad urban leaders: there are flexible and down-to-earth ones and there are radical and sectarian ones. Minimising the number of the latter is essential so the city can advance without becoming paralysed by unnecessary conflicts.

One of the key factors in making headway with an ambitious transformation programme is having as much intelligence available as possible, beyond just your own team. One of the first decisions I took was to form the Urban Habitat Advisory Board and offer Oriol Bohigas the presidency, which he accepted with pleasure. Together with Oriol, some of the members of the multidisciplinary board were Joan Majó, Carlos Ferraté (who later became president of the Board), Ramon Folch, Josep Amat, Mateu Torró, Antoni Font, Enric Massip, Jordi Badia, Anna Cabré, Lluís Comeron, Andreu Veá and so many others. The Board debated the most important projects in the city: from Glòries, some of them were already members of the international competition's judging panel (as you can see here at

> http://ajuntament.barcelona.cat/glories/la-transformacio-de-glories-en-marxa/concurs_explicacio/),

as well as Smart Cities as a concept. The Board never acted as an arm of the municipal government (it would have been impossible if you look at the salaries of some of its members). In fact quite the opposite: it became a space for constructive criticism and for proposing ideas that was highly valued by the entire governmental team. My intention was to bring together a group of professionals and academics who covered the full spectrum of Urban Habitat, taking into account their professional recognition, without any kind of ideological apriorism. I sincerely think that it was a unique experience, and the city should be grateful to the people who so generously gave their time to make it better, and to greatly improve our ability and our skills. This is an example that any city in the world can follow. What's intelligent is relying on the intelligence available, and involving it fully in the city's transformative processes.

8

Participating, Experimenting and Superblocks (Or the City is Anything but Simple)

WINNING ELECTIONS ISN'T CONVINCING ENOUGH

Let's talk about participation. Let's choose a complex, large city. Let's choose one of the best mayors in the world. When we left the mayor of Bogotá's city hall office, we understood that Enrique Peñalosa, a transformative and visionary mayor, had been left a city plagued by problems and debt by his predecessors. We also understood that it would be very difficult for him to convince his community to pick themselves up and break free from the negative spiral to which they felt captive. A few days before he had bitterly argued with an old street vendor who was harshly reproaching him for having thrown them out onto the streets. The video of the peddler chastising a seemingly arrogant mayor who appeared deaf to the woman's requests, circled the world. Peñalosa was shown in the video as something he is not: a cold man oblivious to people's requests, to his people's requests.

It's easy to criticise a mayor. That loneliness at the top is only stomached by the strongest. And it's up to them to take decisions. However, the mayor embodies communities and people, street vendors included. The video revealed that politically disastrous separation between what we are and what we represent. A separation that in these so-called times of post-truth, which I call times of cynicism, seems to mean that the only people that dominate are those who take advantage of the good faith of ordinary people to ramble on, give hope and then leave high and dry the very people that got them to the top. Peñalosa wasn't one of these people and he suffered.

Bogotá is a marvellous and monstrous city. It's very tough and very welcoming. I'm in love with it. The fabric of Bogotá has been woven by the entire history of an America built on an encounter that set discovery and its millions of consequences in motion. An appealing, bloody America, beautifully darned with gold thread and barbed wire. Bogotá is witness to shanty towns, callousness, and banks of social hurt that charge inflationary interest. Bogotá walks to the edge of survival and on the catwalk of opulence. It's the astral distance that separates the Gun Club from Ciudad Bolívar. This is its problem and that of the majority of the modern world's megalopoli. But this doesn't restrict Bogotá from having hope, quite the opposite: Bogotá beats with a heart as passionate as Bolívar's . . . If Simón Bolívar were to raise his head he would advise the mayors of this world to avoid tempting the fate of the seemingly absolutist devil that's inside some of them.

Because the Liberator decided to proclaim himself the dictator of Bogotá, on a rainy August afternoon in 1820, whilst in the neighbouring Teatro Colón a Hungarian dancer performed dances from the Danubian regions. Or was it a soprano from Malaga? What does it matter! What's important is that Bolívar, tired of his mission, and probably of himself, decided only to listen to the voice of his other. And he got it wrong. It was then that the liberal Colombian Socratic Parliamentary Society realised that the monster controlled the General. The liberals, awake amongst their drowsy colleagues, gathered in rooms full of chiaroscuro, perfumed wood and libraries full of Voltaire, Rousseau, Ariosto, Plutarch, Cicero . . . These enlightened men argued heatedly, tried to postulate and arrived at an important conclusion, developed by the playwright Luis Vargas Tejada: *Si de Bolívar la letra con que empieza y aquélla con la que acaba le quitamos, oliva de la paz símbolo hallamos. Esto quiere decir que la cabeza al tirano y los pies cortar debemos si es que una paz durable apetecemos.*[12]

It seems obvious that don Luis' muse had abandoned him that night, probably because a consignment of Venezuelan rum had recently arrived in Bogotá. It is very probable that he also hadn't inter-

[12] The verse literally translates to, "If we remove the first and last letters of Bolívar, we get olive (*oliva* in Spanish) the symbol of peace. This means we should cut the tyrant's head and feet off if we want long-lasting peace."

nalised the admonitions that Shakespeare expresses in Mark Antony's speech about the honesty of Julius Caesar's assassins. What's clear is that the enlightened decided to kill Bolívar. At midnight on the 25th of September the assassins infiltrated the Palacio de San Carlos, without realising that Manuela Sáenz would be waiting for them. Bolívar escaped through a window while his lover distracted the conspirators. Although Bolívar regained control of the situation, nothing would ever be the same as his former dream, and Gran Colombia started to break up until it became what it is today. Listen, listen, listen, the solitary statue of don Simón Bolívar seems to tell us, standing in the centre of the capitals' plaza Mayor. Listening is what everyone who wants to devote themselves to managing cities must do. Listening is the Smartest decision, the most intelligent, the most profitable for the society we serve.

Twenty years ago, when Bogotá was fortunate enough to have the support of visionary mayors like Antanas Mokus or Peñalosa himself, the city advanced. It did so, obviously, in the wake of the 1991 constitutional reform, the first step Colombia had to take to win peace, peace that has been so brilliantly reached today by the persevering intelligence of President Juan Manuel Santos. Fifty years of guerrilla warfare aren't eliminated on the back of one constitutional Article. But it is possible that between the Constitution, political willpower, and the generosity and imagination of those who enter the public sector to work with others, general living conditions can improve and prepare Colombian society for creating present and future opportunities.

The 19th of April Movement (M-19) abandoned the armed struggle and created a new type of politicians in Colombia, who were very close to ordinary people. They seemed to oppose right-wing populists, those who used common people as a way of perpetuating themselves and their social stratum in power. This isn't a Colombian phenomenon, it's universal, although the history of Latin America has undoubtedly contributed to producing some very prominent examples. Gustavo Petro was the mayor before Peñalosa. Petro had been a guerrilla in the 19th of April Movement, and today is one of the most high-profile politicians in Colombia, even after his controversial episode as mayor of Bogotá. Petro, who in his day participated in the taking of Bogotá's Palacio de Justicia, in the plaza de Simón Bolívar (one of the most beautiful squares in the world), became the city's mayor, and occupied the chair that faced the building he had helped to set on fire a few years before. Whilst some think this is a scandal, I like to think of it as a test of maturity for Colombian society.

Achievements are another thing entirely, given that we're discovering that this new type of politics is represented by great popular leaders that speak plainly, who are legitimised by fighting (whether armed or pacific), and sincerely moved by a desire to change the state and the *strata* of things that place old problems within the new political agenda. They make large promises and achieve little. In this way they're the same politicians as always: great objectives often become paralysed by lazy idealism and digressive rhetoric that walks the line between naivety and, a lot of the time, fallaciousness, achieving almost nothing.

It's true that, unlike the most stagnant and dangerous parts of the right, and some red-shirt-wearing leftists, the Petros that we all know in our own countries don't try to reproduce a caste, but try to change things with something other than the conjugation of verbs. What's bad is that leftist populism, as enthusiastic as it is not very effective, produces frustration among people of faith. They lose faith in a better life because they dream about paradise whilst being poorly equipped professionally; they end up bemired in the purgatory of harsh reality.

This was the Enrique Peñalosa that we met in his office, occupied since the 1st of January 2016. It was the 1st of May, precisely the day on which Petro's followers, with him at the fore, were rallying. Enrique seemed to be perplexed, "How can these men demand that I don't sell that useless platform they set up? With the money that we're going to get out of the sale we're going to build more than thirty schools, two hospitals, I will extend the Transmilenio (the system of fast buses that cross the city; the basis of his global fame as a mayor) . . . I'll have to work very hard to make it clear that this is what most benefits the people of Bogotá," he told us, sure of himself but logically worried. He knew that despite being right, it wasn't enough. He needed to produce further legitimacy. Winning elections legitimises the right to direct, to manage, but once they have been won it's advisable to find out what people think in a direct way, in the knowledge that the 21st century's representative democracy must intelligently combine the legitimacies that have come out of the polls with direct knowledge of people's sentiments that can be provided by digital tools. Sentiment, objective data, opinions, votes. Today everything is mixed, and we must separate the wheat from the chaff as quickly as possible.

PARTICIPATING TO ADD VALUE: MICHELANGELO'S DAVID

Lately, in the world of reflection on and management of cities, much has been discussed about direct democracy. Participatory processes have been defended and glorified to exhaustion. But talking about participation isn't talking about something new. To participate is to give an opinion, of course, but above all it is to form part of the process of creating criteria that will lead to decisions that, in the end, must be taken by those who have the legitimacy to do so. We are fortunate that today, well-applied Smart technology allows us to simultaneously talk to every one of our citizens; it allows them to talk amongst themselves and we can carry out a consultation process with everyone at the same time, cutting out any middlemen, adequately diving into segments and with high levels of participation.

The history of participation didn't start the day before yesterday, nor does it pertain to the movements of the latest pseudo-spontaneous generation. I could talk about the process of participation put into practice by the Girona Cathedral's Chapter in the middle of the 14th century to decide how to enlarge it, based on speaking to expert architects and clergymen who, after a number of sessions, decided to go ahead with the project. However, I will stop in Florence, with the abbot of the Cistercian monastery of Clairvaux, in December 1520, on his way to Rome. As he approached the headquarters of the Florentine government in the Piazza della Signoria, he stopped in front of Michelangelo's David, that five-metre tall block of marble. The abbot was surprised by the figure of the young naked man. He wasn't able to recognise the character, nor interpret that its meaning symbolised the strength of the Florentine nation. He simply knew that the statue had been put there after a genuine process of citizen participation that had generated debates and discussions of all kinds. The abbot would have liked to intervene in the controversies aroused in Florence by that uncertain David, to whom they had dedicated so much discussion, and who was undoubtedly a passive protagonist in the first great participatory process of the western world, *alla manera moderna*. The figure seemed ghostly to the abbot, and he never would have put it there. The abbot, however, sensed that the process had been long and arduous. He couldn't say anything.

Michelangelo's statue was originally intended for the cathedral, destined for a position on top of one of the buttresses in the so-called Brunelleschi's dome, at a safe distance from the population. It was designed as part of a series of sculptures dedicated to the prophets that were to be installed on the cathedral's buttresses. Years before, terracotta figures had been positioned on two of the buttresses, Joshua by Donatello and Hercules by Agostino di Duccio, painted white to look like marble. Whilst working on his David in the works area of the cathedral, just below the dome, Michelangelo could study these models. If David's head looks too big for his body, or his facial features seem too exaggerated, we have to imagine that they were meant to be seen from below and at a great distance, or rather we should imagine Michelangelo trying to imagine it.

Michelangelo's work was not only conditioned by its originally planned location, but also by the fact that the block of marble that was given to him had been exposed to the elements since the 1460s, and it had been worked on, perhaps extensively, by Antonio Rossellino in the 1470s. On the 2nd of July 1501, the members of the cathedral's works council agreed to commission a marble statue, which they would name David, from the block located at the rear part of the aforementioned works area. The agreement indicated that the block should be lifted and stood up in order to assess the feasibility of completing the sculpture. Six weeks later, the 26-year-old Michelangelo Buonarroti received the commission. At the start of 1502 the giant, as it is called in the documents, was half finished. In June 1503 the statue was complete enough to open up the sculptor's work space to anyone who wished to see it, an occasion that coincided with the Feast Day of St. John the Baptist, patron saint of Florence, when the city exhibits its richest treasures.

For Vasari, David's extraction from a mutilated block of marble was nothing less than a miracle; Michelangelo had managed, "to bring back to life one who was dead." From that day on the public discussion started about where the young sculptor's statue should be placed. The majority's opinion was to find a new place for it to bring it closer to the street. There are no other documents on Renaissance art similar to the record of the deliberating of the group of experts that convened on the 25th of January 1504, to decide where to put the statue. Practically all of Florence's great artists and architects, like Leonardo and Botticelli, were summoned (in great haste, as it seems) to intervene in a seemingly astonishing discussion, already turned into a state issue.

Citizens of the city's social strata were invited along with the artists, in what was an authentic participatory process that was well carried out and even better documented. It's important to note that, after the Medici's expulsion from Florence, people had become used to hearing that no citizen should ever allow anyone else to rise above the rest. For this reason Florence had entered into a period of great public debates, between them, the statue and its placement. This was happening in a city struggling to regain its communal republican traditions, at a time of foreign invasions. David, in that sense, acquired the character of a symbolic lightning rod, in the middle of the era's identity crisis.

According to the minutes taken in the meeting, only the carpenter Francesco Mocciatto wondered why they wouldn't put it in the place in which it been designed to go. In fact, Mocciatto complained because, he assured, that whatever was decided for the project, the authorities had already decided to put the giant on the street. The other members of the committee weren't entirely sure about where to position David. All of them understood, however, the need to place it at street level, with the safeguard that the statue was three times larger than a person, something that could severely condition the traffic in the street it was located. Questions were repeatedly asked about the effect the statue could have on pedestrians. In the same way, they valued a placement where the statue could be seen from different perspectives.

Of all the opinions expressed in the meeting, the most striking one was offered by the goldsmith Andrea de Riccio. He thought that it should be placed inside the Loggia dei Lanzi, next to the Palazzo della Signoria, not only to protect it from the elements, but citizens from it as well. The lodge would have framed and imposed a frontal view of David, especially if the architect Giuliano da Sangallo's suggestion that it should go in a niche at that back of the lodge had been taken into account. Finally they decided to place David in front of the Palazzo della Signoria, to the side of the main door, where it would replace Donatello's statue of Judith. This statue, that had belonged to the Medici until they were exiled, and installed in front of the seat of government in 1495, had never been loved by Florentines. The head of the Florentine government, the first to speak at the meeting on David's placement, insisted on the removal of Judith. He called it, "*segno mortifero*," emblem of death, and declared, referring to the theme of Donatello's sculpture, "It is not befitting that a woman should kill a man, above all this statue was erected under an evil constellation, we have gone continually from bad to worse since then." One might think that a man who believed in effigies with deadly powers and evil astral conjunctions would have been perfectly happy avoiding any kind of deal that involved a statue. However, in Renaissance Florence, art was the solution to the problem, even when the problem was another sculpture. The victorious David would replace Judith to pave the way for a new age in troubled Florence.

Taking this logic into account, there would have been no moment more dangerous in David's management than when he was in transit, freed from the works of the cathedral but still not erected in his new location. On the 14th of May 1504, at eight in the evening, David, upright in a carriage, started his journey from the cathedral to the seat of government, pushed by more than forty men. The process took four days. The permanent placement of the statue in front of the Palazzo della Signoria earned the general approval of the city from the first day. The Florentines, thanks to the preceding process of discussion, had made it their own. They, prone to inventiveness, made up all kinds of fantastic interpretations about the giant. What is evident is that from the outset, the effect sought by the leaders of the city had been achieved and David, beyond symbolising the strength of Florence before its enemies, ended up becoming the Florentines' very identity.

CIVIC APPROPRIATION

A participatory process for civic appropriation. This is the great hidden objective of the intelligent Florentine patricians, that they sought in the symbol and in the process of legitimisation, a way of uniting the populace around a shared premise. Today, however, we can see how many participatory processes serve only to polarise. The long process David's relocation was submitted to demonstrates the extent to which the participatory process can be useful to legitimise a decision that, in spite of the fact that after centuries it seems obvious, at the time it wasn't so clear. On the other hand, it's important to highlight the extent to which participatory processes cannot be considered a new democratic idea, as it is often assumed. It's true that the Florentine Republic was a mature political unit, with experience of dictatorial governments and experience of having suffered the auto sacramentals of the purgative Savonarola. Florence was, therefore, vaccinated from the despotism of the enriched and enlightened figures of the Renaissance; and from the despotism of those who believed they had an inquisitive moral superiority. In this Florence, and by extension, Italy, became a precursor again, dedicating time and effort to consolidating the decision taken by the upper echelons of the Florentine Republic: to descend from the cathedral's lofty heights, which they perceived as the symbol of the city's new era, and why not, an entire culture's. Herein lies the greatness of operative, consultative participation that generates criteria and consensus based on a vision. Participation orientated towards doing, towards defining criteria, but not towards blocking or substituting representative chambers' democratic legitimacy. Intelligent participation, that is to say, open to using the available technological tools, and with them the strongest possible arguments.

The real challenge of participation is therefore to achieve this direct relationship with citizens, which allows us to move forward without leaving anyone behind, but without being conditioned by those who shout the loudest or take hold of the assembly. This has already been proved in the country of the soviets, or in those of organic democracy. Knowing how to synthesise the criteria that must determine the what and how of a city is crucial, but I'm absolutely against the transfer of technical responsibility to those who neither have the preparation, nor the necessary knowledge. Put more clearly, elected politicians have to decide on what and where they spend their budget. That is to say, they have to decide what has to be done, given that this is the reason they have been chosen. A political player has to base their decision on their personal criteria and permanently listening to the citizenship they serve. To date, surveys have done this job. Not long ago, incentivised by politicians with whom we share a way of understanding public service, technological platforms for citizen consultation were developed; they allow these criteria and their evolution to be visible. In the end, it is the experts who must translate the solution into a physical reality for the city. The direct relationship with citizens, therefore, doesn't have to pass through any kind of mediator unless it is strictly necessary. However, in many cases it takes excellent civic representation with which to debate complex and concrete problems. Civic representation is a social asset that must be preserved and enhanced, without any kind of nuance. When I refer to mediation I am talking about those who, by manipulating the opinion of the majority, speak for the few.

The challenge that is bearing down on us is learning how to stimulate creative and informative participation that must come directly from our neighbours. How to convince them that their opinion is being effectively listened to and that its value is more than that of a participatory placebo. In Barcelona we made a mistake at the start of our mandate. We couldn't wait to start! We retrained our sights on la Rambla de Sants and in Glòries, in a different way in each case, always aiming to positively transform the neighbourhoods and the city. Later on, we made a mistake in failing to adequately gauge deep-rooted opinions in some areas of the city. Neither were we able to understand that some neighbourhood mediators were nothing more than conveyor belts of the old renewed policy of opposition of small groups. Barcelona knows the extent to which neighbourhood leaders with records of impeccable service up to a certain point in their lives, have ended up as directors, bar owners, people in charge of maintenance, in a civic centre or facility . . . This appropriation, this quasi-privatisation of the civic, typical of totalitarian ideologies, or its offspring camouflaged as all kinds of populism, that has been so in vogue lately. Social life must end up revolving around those who decide what is good and what is bad. The superiority of the party and its morals end up justifying everything, since the party and its ideology are the only acceptable glue that holds society together.

PARTICIPATION: SUCCESSES AND ERRORS

Unquestionably, without great social sensitivity, without the bravery to test all the evidence, some working-class districts of Barcelona would never have escaped their impasse. It's true, as Marc Andreu (a Catalan historian) explains in his thesis on neighbourhood associationism in the city, that an important part of the driving force for change in the latter stages of Franco's regime sprang from this movement. However, its resurrection in the heart of the municipal administration, converted into a parallel administration through long-term agreements that end up being perceived as law, in many cases culminates in throttling the city's capacity for change in key areas, principally in neighbourhoods.

An illustrative example of the consequences of this institutionalised lack of perspective was the unbearable demolition of the wall that separates Ciutadella de Barcelona park from the Vila Olímpica district. This wall is one of the most absurd things our city has faced, even more so since the installation of the Universität Pompeu Fabra's headquarters in the old barracks of Jaume I. In fact, one of the challenges we worked on was designing Passeig dels Til·lers' route over Estació de França's train tracks towards the sea, as well as getting past Ciutadella's wall, with the aim of connecting Poble Nou and everything that is emerging in the east of the city, with the historic centre. These points of connection would have turned Ciutadella park into a high-quality hub of intergenerational knowledge: primary and secondary schools, institutes, universities, professional training, Master's degrees, postgraduate studies . . . everything around the park and the Parliament. Ciutadella at last turned into a great distributor of social and urban energy, and what's more important, into a distributor with a heart.

We mulled over the Ciutadella's real purpose in the 21st century. A constant inspiration for this movement was the continuous conversation we held with Professor Mas-Colell, father of the Universität Pompeu Fabra, of the Barcelona Graduate School of Economics, and one of the most magnificent urbanists I have ever known. Andreu Mas-Colell, observing Ciutadella from the Barcelona GSE building, helped me to see beyond the urban project, to find the motives in an authentic civic project, reproducible even in other cities with similar problems. Willy Müller and his Barcelona Regional team, in collaboration with María Sisternas and Project Management, got to work.

The presence of the Parlament de Catalunya gives the place an authentic institutional seal, but it's also true that, in the best Catalan institutional tradition, in which seats of government and representation are mixed with daily life, in Ciutadella one can feel the city's pulse by listening to the children's voices playing in the courtyards of Parc primary school. One can feel it sitting next to adolescents from the Institut Verdaguer, or walking next to Bachelor's or Doctorate students from the Universität Pompeu Fabra. In other words, Ciutadella, without us realising it, had become the first intergenerational campus in the world, in the historic heart of Barcelona. Furthermore, it had the capacity to interconnect the activities of the Hospital del Mar, the Biomedical Research Centre, the urban power of the city's historic neighbourhoods originating in the commitment to 22@,[13] and the connections to Eixample. Following some of Solà Morales' lessons, I realised to the need to open up, to connect in a seemingly smaller point of the city, but, like in that children's tale where a seed is thrown to the ground, that would allow us to grow a *beanstalk* to take us to the sky. Others had tried it, but without giving the key idea a chance: generationally and physically interconnected education.

We sat down with the university, the secondary school and the primary school; we opened an entrance to the primary school on the same side as the zoo, we achieved the parents' consensus, we developed the project in accordance with the Parliament's security needs, negotiating the country's entire institutional hierarchy. Who could oppose such an objectively good initiative for everyone? Opening Ciutadella on Carrer de Wellington meant opening up the city in a key place, subjected to a lot of social and economic pressure, currently creating inequality and limiting opportunity. We opened up Wellington by moving the dolphinarium to another area of the park and by redefining some of the zoo's spaces. Despite that fact I was, am and will be in favour of moving a substantial part of Barcelona Zoo to land on the agricultural park in Barcelona's metropolitan area, near the Llobregat river. To this end we established a good transactional model with the always tough representatives of the professional unions within the Zoo. The project foresaw a path for pedestrians at ground level, while the Zoo would rise above this path using a footbridge converted into an ecosystem for birds. It was an imaginative solution to the problem of the coexistence of the zoo and people's freedom of movement. It was also a first attempt to change the dimensions of this coexistence using common sense and giving the necessary importance to the connection between people, in relation to the zoo.

The project failed. They made it fail. I failed, and the city with me. I realised the enormous need to legitimise the action taken in the city in a different way, and the only way of doing so is to constantly collaborate with the maximum number of citizens possible. The key question, however, is: when to do so? When and in what way is it necessary to define this bijective collaboration? At the time of defining

---

[13] The name given to an area of the formerly industrial Poblenou that has been urbanistically renovated.

the project criteria? At the time of creating the pitch? At the most open moment of the conceptualisation and design phase? Technology allows to do so in two ways: in person or electronically. We failed in Ciutadella because we were unable to see that we could have taken advantage of its number of visitors. The zoo, the park, the educative facilities, the population of Vila Olímpica should have made us see the possibility of trying processes that involved a direct relationship with people based on statements on one side of the wall and the other, on assessments and *in situ* analysis. On the other hand, we should have had to install platforms that would have allowed us to obtain opinions and reflections from people at home. The balance between the opinions of nearby residents and people from the rest of the city would have given a double result: on the one hand, we would have improved the project; on the other, we would have compensated for some people's obstructive capacity. We didn't do it. The project remained in a drawer.

In short, here the necessity for the development of consultation and participation platforms through Smart technology comes into play. Because without building a direct relationship with citizens, there will always be professional representatives of public things, connected to no-to-everything politics, that are going to halt the city's progress. It's a clash of legitimacies that must be overcome with objectivity since these people are particularly adept at managing the subjective, making it appear objective. This is the biggest problem, in fact the biggest danger, that must be faced by representative democracy: to legitimise itself through things that are obvious, that is, only this democracy is subjected to the objective scrutiny of votes. Although technology does not free us from all pathologies, it can, in this case, reinforce this legitimacy by building an active community of collaborators for public good, expressing constructive opinions, collaborating in its development, and avoiding becoming a conveyor belt for low-flying political struggle.

SUPERBLOCKS, OR HOW TO SEE AND LISTEN TO THE CITY IN A DIFFERENT WAY

The launch of superblocks, worked on with this philosophy, was a success, however. It was very important from the point of view of public action, and very discreet from the point of view of public exploitation. We wanted it this way, precisely to guarantee success. The key? The silent way we work in depth, participation as a base for the structure of its application and development criteria, and the mass use of data through Smart technologies. In short, its application to the design of the urban transformation of a city as mature as Barcelona, in which a historic debate exists on its mobility model, with very important implications for people's daily lives.

Cerdà's grid system is an isotropic blueprint (all the streets and blocks are as wide as each other) but since its origin it was conceived as a superblock model that favours some roads for higher amounts of vehicular transit, leaving others for more reduced mobility, on foot. The new bus network and 2013 the Mobility Plan already consolidated the most complex model, that which set out hierarchies of roads every three blocks. But in our opinion mobility is no more than one of Urban Habitat's layers. We implemented the Superblocks Programme using a cross-discipline perspective incorporating the five pillars of Urban Habitat (urbanism, environment, housing, infrastructure and IT). We defined, for every field, a set of actions ranging from the rehabilitation of buildings to Smart management, with the aim of achieving active regeneration of the urban fabric through innovative projects defined by the residents or users themselves.

The superblocks project itself is atypical of conventional urban projects. In fact, it's a programme (not a project) that starts with a technical diagnosis, but trusts and depends on civic implication to design the road map that needs to change public space, improve its appropriation and increase the co-responsibility of all its representatives on the road to an intelligent city.

The superblocks that we designed were based on the Barcelona mantra: productive neighbourhoods, human speed, energy self-sufficiency, connectivity, zero emissions, quality of life, equity, culture and inclusiveness. The key to making this happen is that cities, businesses large and small, and schools and universities that give professional training, needed to learn how to work together to turn this vision into commodities of basic urban services, in exactly the same way as the generalisation of connection to electricity, gas and water grids, as well as the spread of telephone lines, transformed the country in the age of the Commonwealth of Catalonia, at the start of the 20th century. This is our generation's challenge and we are committed to it.

Superblocks must be promoted in parallel with the city's large transformation projects, that have to build more affordable housing and new urban centres that have been unimagined up to now. Only through the combination of both scales (internal regeneration, block to block, testing innovative formulas, and the consolidation of new, denser and well-connected centres) will citizens be able to combat urban obsolescence, the tendency for housing bubbles and a lack of equity.

How did I do it in Barcelona and how do I do it in other world cities that I now work for? Where did I start? It's very simple: looking at the city through stories, through the million things that a city is, as they used to say on the radio in the seventies. Josep Pla (a contemporary Spanish artist) didn't like those flats in Barcelona's Eixample with interminable corridors and interior rooms. Any foreigner, to get to know Barcelona, has to have spent a few days living in one of Ensanche's flats. By doing so they'll understand that flats in the centre are very long and thin, with one façade facing the street and another the block's interior courtyard, and with three or four rooms whose only ventilation are these exiguous courtyards we call *cel-oberts*. The majority of the flats in Ensanche are very similar to each other, with high ceilings, large rooms of a similar size with long corridors that wrap around the building's staircases. They're flats in buildings built by the bourgeois of the 19th century whose heirs have kept the properties. They are not the Victorian houses of London, nor the apartments of Paris' XVIème; Barcelona is a far less wealthy city and this shows in its housing stock. However, despite Pla's opinion, they are well-built flats, with well-proportioned rooms and façades that look out onto leafy streets, with large windows that let sunlight in. Ensanche's luxury is the following: craftsmen were employed to mould ceilings and design magnificent tile floors, in flats that can be lived in, but can also be used as an office, study, workshop, surgery or architectural firm. Ensanche is probably the district that best embodies Barcelona's fame, "One lives and works well."

Julia Fournier Cuadros was born in 1887. She was a privileged woman. She lived on the main floor of Carrer de València 261. Her father had the building constructed in 1896, on the corner of Passatge dels Camps Elisis, which at that moment in time had to have been one of the most select places to live in the city, with the terrace of the Gran Hotel Majestic opposite. The capitals that support the mezzanines' sills, sculpted as "heads of colonial slaves," are a testament to this. Mrs Fournier studied at the Col·legi Sagrat Cor de Sarrià, as elitist then as it is now, and she spent her entire life moving between Barcelona and La Garriga, where her husband accumulated property. At the end of the seventies, Julieta, as she was known by the neighbours in her building, was no longer the bourgeois lady that she should have been, but an endearing old lady with a very fragile appearance, who strolled daily across the main terrace, with a linen umbrella, accompanied by some of her assistants. Maria grew up in one of those flats during the eighties. Her uncle chauffeured Julieta to pay for his degree. When the uncle explained to Mrs Fournier that his brother had a little girl and he was looking for a flat, Mrs Fournier offered them all the empty flats in the building. It was the eighties and Ensanche was still a district that needed to be rediscovered. Ensanche was deserted. Julieta made it easy for that young couple: they could choose the flat that they wanted to rent, and agreed a more than affordable price with an indefinite contract. They chose the 3º 1ª, which wasn't the best because it didn't look out onto the Passatge dels Camps Elisis, but it was the one that needed the least doing up to be able to move in immediately.

Maria grew up in that flat, with its good things and bad, watching the minuscule black and white television in her bedroom, indifferent to what happened in the dining room in which her parents tried to get by. In summer they had an inflatable plastic pool in the porch that leaked water all over the tile floor. It was unbearably hot in that porch, but the spilt water was refreshing and it made the passing of hours more pleasant. Ensanche's porches are full of everything: butterfly chairs, tropical hammocks, genuine domestic zoos made up of aquariums, aviaries and cages in which hamsters metaphorise us humans as they run on their infinite wheels. These porches are the playing field for beginner skaters, stages for nativity scenes and Christmas trees, as well as a corner for sewing and collections of comics.

Maria told me that she began to study architecture in that "independent" study that was the porch of her parents' flat in Ensanche, from which she could hear the television in the living room, whilst she traced kilometres of parchment paper with *Rotring* pens, precursors to digital life. The porch was above all the house's official *smoking room*, where the adults consumed cigarette after cigarette whilst they spied on all the other porches on the interior of the block. But these porches were above all places of tranquility, averse to the noisy extremes of the street. On the other side of the house the carpentry shook loudly as cars and buses went past. The street entered it in the shape of stag dos that lasted until the small hours of the morning, arguments between deliverymen and pedestrians in a hurry, or simply the seemingly insignificant, informal conversations that form the soul of the city. The block's interior courtyard, however, was pure peace. We Barcelonians have mentally photographed them so many times . . .

There are fewer and fewer flats in Ensanche that look out onto the street and also onto the block's interior courtyard. Many of them have thermal, acoustic and even lighting problems (there are many interior rooms that don't receive any sunlight throughout the entire day). Some of the blocks' interiors have been expropriated and became green zones in the city's General Plan. Others are completely private and are not accessible from the street. But the palm trees, the overflowing trees and the bougainvilleas covered in flowers, the Catalan terraces that inhabit the interiors of blocks, and their peace, signal common good that benefits the community that lives in the block. In a city as compact as Barcelona, these spaces of decompression are a major factor in quality of life.

In the conversations prior to the launch of the superblocks project, Maria and I spoke often of the need to strengthen this balance. We were convinced, and we still are, that we can improve quality of life without penalising some people to the detriment of others, without condemning the city to having first and second class streets; without renouncing the idea of social justice, equity and isotropy that forms the fabric of Ensanche.

With the data in hand, Barcelona has an exceptionally high density. Barcelona's mobility model is excellent, compared to the traffic jams that are collapsing the globe. There is a high modal distribution in favour of public transport. Although a large number of people own cars, they are mostly used only when the alternative to public transport is not feasible. Having a denser metro network would be desirable, but it has been years since there have been traffic jams in Barcelona that threaten liveability, as is the case in many other cities. There are ambitious mobility plans to improve the connection between districts, such as making Carrer de Balmes a one-way street, that according to some experts would facilitate the recovery of a decent quantity of the pavements above Plaça Molina. In any case, the large connectors must be fluid so that the interior streets experience reasonable reductions. As with everything, the best thing is gradual change, which allows users to change their transport preferences in an evolutionary way.

Residents of Ensanche rarely use their cars for their daily journeys, because it's the district that's best served by public transport. Furthermore, the system is so isotropic, thanks to the grid system, that a problem at any point is solved by ramifying the traffic through the parallel streets without the need to intervene. Closing transit in Ensanche means putting more pressure on the rest of the districts, with the danger of turning it into a museum-like city, to the detriment of the poorly-named peripheral districts. As a result, the superblock model had not been implemented before, although the term was coined in the thirties, developed in a conceptual way by Professor Rueda and his team, and finally brought up to date through the contributions of the Urban Habitat team.

Convinced that we needed to scale up the application of the concept, we decided to approve an ambitious and equitable Government Measure on superblocks for the entire city, beyond Ensanche. Furthermore, we determined that the superblocks strategy did not merely serve to pedestrianise the centre, but that it was an innovative programme for the regeneration of the city from the inside, promoting all those ground-breaking solutions that the General Plan had prevented until that moment. A cross-discipline Urban Habitat programme that covered all possible dimensions: energy, services, biodiversity, economic activity and mobility.

We devised it as a programme capable of providing, in the selected areas, a high quality of life, with solutions proposed by the citizens themselves or by the administration. The ideas we put forward included solutions like these:

1. To maximise the interiors of blocks as green and silent spaces, re-naturalised.
2. To create spaces with sun and shade, both are equally necessary in a mediterranean city.
3. To resolve innovative projects with technical solvency and excellence, all those challenges that citizens raise in participatory sessions: to put knowledge at the service of ambitious goals.
4. To rationalise mobility whilst maintaining homogeneity throughout the city (congestion loads distributed across the grid, instead of first and second class streets).

5. To use the interiors of blocks as spaces for the accumulation of renewable energy.
6. To move the noisiest, collective leisure activities to reenergised green terraces.

Superblocks are not merely a question of urban technology, but an element of the city's cultural and identity configuration, a tool for the reconfiguration of citizens' perceptions of public space and their interaction with private domestic and civic space. A superblock integrates urban services: water, electricity, logistics, recycling, information, social services, combined with all dimensions, that is to say, on a housing level, building, street, square . . . We did it, therefore, by unpackaging the concept and redefining it in favour of distributed energy, in favour of the promotion of the ground floors of buildings as new centres of production. We did it by making the most of free spaces to develop urban acupunctures, optimising the investment and maximising its social impact. We did it without stigmatising private vehicles, but by inviting people to use them in a different way, that is, to use them less, with the help of the sensorisation of the city.

We did it in silence, but everything's on the internet. You can see it at

http://www.sostenibilitatbcn.cat/attachments/ article/1233/ Dossier%20 Descripcio_web.pdf,

OR AT

http://ajuntament.barcelona.cat/superilles/ca/superilla/les-corts.

There we developed the strategy and the first tactical movements; at the same time we began an intense plan of workshops with residents and the principal co-ordinators of the areas in which we started to work. As always, it was essential to anticipate resistance to change. We managed to abandon our own and other people's prejudices through frank conversation with as many people as we physically could. The Les Corts superblock is a prototype because it is not in a neighbourhood with high purchasing power, nor does it have large or elegant façades; and it has endemic problems with starting-up businesses on the buildings' ground floors. In fact it's an exemplary superblock from the point of view of urban problems. Anna Ramon, the president of the Camp Nou residents' association, opened my eyes to this hidden reality that are the streets behind the city's great landmarks. It's an eminently residential area that suffers avalanches of people on football days without sharing any positive externality from this overwhelming presence. How to regenerate the place? How to give a better quality of life to a consolidated area of Barcelona that does not stand out for any reasons, that only exists for its residents? Ana and her arguments served to shape some of the proposals for the new Camp Nou, and in fact it was Bjarke Ingels and the BIG team that spoke most about this integration, in which the coexistence of stadium and city of mixed use, should be strengthened. Bjarke incorporated the Camp Nou in the superblock to put it at the service of the city. His was not the winning proposal, but I'm sure that this concept of urban, social and economic integration will bear fruit in any of the architectural proposals for the new stadium.

During our continuous walk through the city, to have time to think alone, something we will never renounce, to reflect, to interpret and thus to be able to discuss and act, we realised that the approach was perfect for the Hospital Clinic. Barcelona's massive university hospital, international centre of reference, has been struggling for years to become a definitive place in the new Barcelona. What's certain is that the opportunity was right there, next door, in l'Escola Industrial, the old manufacturing facility founded in 1869 and transformed in 1904 into the Escola del Treball[14] thanks to the genius of the first Catalanists of the 20th century. This is Barcelona's real golden block: industry, education, work, science, health, university, heritage, residents, centrality. The Clinic superblock has everything it needs to become an authentic world centre for science and education, with an evident positive urban impact. We had to fight against the short-term vision of those who wanted to convert the old faculties into offices for the civil servants of the Diputació de Barcelona. It's not that the civil servants don't deserve good spaces to work in, but it's incredibly clear that the opportunity to grow the university, the faculty of medicine, the Clinic and its ecosystem of startups far surpassed the civil service option. With the addition that this part of Barcelona's Ensanche needs urban regeneration, a physical transformation of the environment that favours quality of life and

[14] The School of Work, a post-obligatory education centre that offers education, training and work placements.

enjoyment of public space is a must. It's in places like this that the superblocks programme makes sense.

In short, the original commitment to superblocks aims to consider all dimensions of the urban habitat, from the subsoil to the terraces, integrating 21st century technology. It's a programme for urban and democratic regeneration in working order, whose objective is to propose innovative and viable alternatives to the old General Plan of 1977, still very important but so outdated and damaged in many of its fundamental aspects. In the Metropolitan Area, a very good team of professionals supervised by the undisputed Dr Joan Busquets has decided to draft an Urbanistic Master Plan for the entire metropolis. Superblocks could be a good testing ground to design new solutions to old problems, above all if they continue counting on the people.

9

Transforming for the Next Three Generations, Counting on Early Victories

THE NEW CAMP NOU, THREE GENERATIONS AFTER MESSI

I'm a Barça fan. It's one of the best things I inherited from my parents and grandparents. One of the few things I have kept from one of my grandparents is a Barcelona crest that my mother's father gave to me just before his death, over forty years ago. Another is a memory I didn't experience but have heard about a thousand and one times at home, when my grandfather took my father to the inauguration (24 September 1957) of the Camp Nou (New Stadium), a replacement for the old "Les Corts" stadium that had been inaugurated in 1922.

Lots of people don't understand why we like football. We like it and we wonder why. We like football and sport in general because of the personal and shared emotions it creates, because it's an education in humility, in life (with all the good and bad that life involves); because it's a social event. For the joy that its victories give us free of charge, and because it demands intelligently controlling the bad mood that envelops you when your team loses. What I like about it are the strategies in offices and changing-rooms, and the tactics on its pitches. I like how much it entertains us, because its myths endure far longer than one moment's irreplaceable emotion, and because it connects us with the examples set and the profoundest emotions felt by those who have preceded us. Haven't you ever wondered what really excited your parents and grandparents, without any intellectual involvement, or the contempt that surfaces in so many snooty views of football in certain environments? I'm lucky enough to know for sure, at least as far as Barça is concerned. Because there's far more intelligence and sincerity, whatever people might say, on a football pitch, than in many other walks of life. I'm a little tired of those who insult me whilst pulling faces, saying they prefer to watch alternative cinema to a Champions League final. I've seen Cruyff fly, Maradona dribble, Schuster pass, Stoickov leave people in the dust, Xavi run the team and Messi turn goals into works of art. I've seen Migueli command, Rifé run down the line, Asensi and Víctor put the team on their backs, and saves from Sadurní, Artola, Zubi and Urruti, of course. Having seen and lived all of this is priceless. Football, and Barça even more so, is a permanent metaphor for life: passion, intelligence, group, individual, glory and letdowns, ideas, practicality, frustration and hope, technical ability, emotional capacity, culture, values, principles (or a lack of them), effort, merit, luck . . . victory and loss. Life.

Thus, we consider ourselves some of the most fortunate people in the world when our time in Barcelona City Council coincided with the club's intention to build a new stadium. President Joan Laporta (President, FC Barcelona, 2003–10) had decided to carry out a remodelling of the Camp Nou, and to do so created an international competition which was won by Norman Foster. Regardless of how much we liked the project, we didn't think it was great, the relevant question was how they were going to pay for the stadium: through rezoning. Clubs like Espanyol and Real Madrid had done it like that recently. Barcelona had tried, but in the end everything fizzled out due to a mixture of political and urbanistic reasons. All in all, when it was up to us to deal with the issue we had the feeling that the city would lose something with Foster's project. Did it make sense to miss the opportunity the club had given us to use the new Camp Nou to bring the city together, to update it, to undo old errors or simply adjust the details?

The Camp Nou was built about one thousand metres from the old Les Corts stadium. The move was the result of Kubala's arrival at Barcelona. The Barça of the Five Cups, the Barcelona that had it not been for Francoist intervention, would have already been in the 50s what it became at the turn of the century: the best club in the world in terms of fanbase, accumulation of talent and way of playing. Well, the Camp Nou was the result of that legendary team formed of Kubala, Ramallets, César and so many others. The rezoning of the Les Corts stadium made it possible to build the Mitjans stadium on the back of, it has to be said, massive debt that led the club to selling Luís Suárez (the old one) and to not winning anything until Johan Cruyff's arrival in 1973. It's true that amongst the club' supporters the construction of a new stadium caused such an impact on the club's members that it's wake can still be felt today, almost in the same way the fear of inflation affects German society, afflicted by the now blurry memory of Weimar.

No, Barcelona shouldn't rezone. The generation of Messi, Iniesta, Xavi, Busquets, Piqué, Puyol and company was, is, sufficiently strong and different to make us think of a economically sustainable, socially useful solution that is orientated towards the future. This was when we saw the opportunity to build the New Camp Nou on land belonging to the Universität de Barcelona located on the edge of the city, bordering Hospitalet de Llobregat. The land, which was used for physical activity at the university, was big enough to house the new stadium as well as the new Palau Blaugrana.[15] It's land that is authorised to be built on, so the exchange between the Universität de Barcelona and Fútbol Club Barcelona was a clean, simple and transparent operation.

Let's pause here for a moment. When talking about Smart Cities, one tends to think that everything is about technology, and this is partly true, although not entirely. In the end, everything ends up with an urban transformation project, because that's what counts: how we transform the space that citizens share so that it belongs more to us, more to everyone, more open and less encrypted. Having football stadium in the middle of the city is no trifle. It can be a problem or it can be an opportunity. Madrid could teach us a thing or two about this, even if it has been thanks to the passing of time. The Estadio de Chamartín was built almost on top of the old stadium where Madrid played, which was demolished to build the current one, in accordance with the extension of Avenida de la Castellana to Cuatro Caminos. In that postwar Madrid, Nuevos Ministerios, Raimundo Fernández Villaverde, Viso and all the districts beyond them, were the peripheries. Over the years the Bernabéu has become one of the world's great stadiums and one of the best integrated in its urban fabric. So much so that some of its surrounding street-corners are among the highest-valued in the city. Madrid demonstrates that urban integration is

15 An arena that is used for basketball, handball, roller hockey and futsal.

possible if citizens' viewpoints, not only urbanistic ones, are sufficiently flexible. Stadiums, Smart Stadiums, are already the most technologically-advanced spaces in a city: what happens on the pitch of a Madrid game is seen by half the world, live, and from dozens of different angles. But it's also happening on people's phones: statistics, replays, analysis, competitions, betting, debates . . . The challenge, for clubs like Barcelona, Manchester City or Real Madrid is making the stadium Smart to blur the liminality between those who watch the game at the stadium and those who watch it at home. Given the alternative of staying at home and watching it on television, the experienced lived in the stadium should be made to be unforgettable.

Large sports facilities must be used with intelligence given that they don't always work in the right way. Let's make a stop in Manchester, another of world football's capitals, where it's possible to enjoy the contrast between Josep Guardiola and José Mourinho's football philosophies again. In Manchester, I'm with City. I like the history of the club founded by a parish priest who wanted to impart positive values on the children in the neighbourhood that came close to the parish. City play in the Etihad Stadium, a stadium built with a purpose, for the 2002 Commonwealth Games. The stadium is on the east side of the city, following the trail marked by the Ashton Canal. I said with a purpose because those who planned its construction wanted to fill the space between the centre of Manchester and Eastlands, and they thought that with the stadium operational, everything would be different. The arrival of Arab money accelerated everything at the end of the century's first decade. Behind everything lies a very special city commissioner, the Labour Party member Sir Howard Bernstein, the champion of modern Manchester, a man who understands that private money, well used, pays a lot of public dividends. Manchester developed this sector of the city, near the stadium, through the investment in one of the most modern sporting cities in the world. Arguments from the era of the Games were reinforced by Manchester City's commitment to converting thousands of square metres of old, contaminated, industrial land into a unique space for elite sport. The problem: being elite meant very few people benefit from the advantages of being there. Therefore the club, and the city, invested in a health centre and a school which unfortunately has difficulty attracting customers. Today Manchester is England's undisputed second city, and it is powerfully making its way to replacing London in some of the services that the capital offered to the world until recently.

What to do with the Etihad? At the end of the day we're talking about the opposite example to the Bernabéu: a magnificent stadium, next to the Ashton Canal, with many hectares of free space adjacent to it, that has only just started to get off the ground. Sir Howard was clear in his ideas: come up with a plan with a solid narrative that would allow productive spirit to return to that half-abandoned part of the city. To do so we proposed filming the days there was a game on and the days there wasn't. In this simple and explicit way it was possible to interpret the place's makeup and, above all, understand the use of the transport corridors that run from the centre of Manchester to the foot of the stadium. Subsequently the team focussed on the old gas storage tanks that neighboured it which were to be given to emerging teams of *makers*. I'm referring to the shining-lights of distributed production, users of 3D printers, those who turn other people's problems into opportunities and make things from a cup to a house, or a chip or a menu. Suddenly we had a stance: Manchester, the capital of the first industrial revolution, could become the capital of the revolution 5.0.

We proposed advancing in stages, with the aim of allowing the city to reanalyse those disused properties that separated Eastlands from the centre. That is to say, a process of gradual and planned colonisation of the empty space between the compact centre and the sporting world of City. It was in our interests to invest in activity surrounding the sporting world and the stadium, to get rid of its intermittent and restricted use, to turn it into a hub of continuous citizen activity, and, once consolidated, into a space that would act as a counterweight for the centre of Manchester, allowing the corridor effect.

The civic development operation around the Etihad Stadium must respond to a vision aimed two and a half generations in the future. The commitment to digital economy as a central argument of the urban and civic operation sets the starting point for what I think should happen in Barcelona. Because the empty space around the stadium must be thought of in terms of what cities are made of today. I don't mean physically, but what they are made of from the point of view of people's expectations.

In Barcelona we had the opportunity to convert the current block of land owned by Barça, in the best urban university campus in Europe, and make the new Camp Nou the regenerative fabric that the heart of Barcelona's western metropolitan area needs. We were building the city and the club in a single operation. We needed to be bold, it's true, as well as a vision for the country and a vision for the club. For once, there was plenty of the former in the political arena, the implicated metropolitan mayors immediately saw the benefits of the operation. What was lacking was the latter: we needed to make the club aware that it was the right time to make a move similar to the one made by their predecessors in the fifties, this time with the difference that instead of building the new stadium in the middle of nowhere, it would be located in the new heart the Barcelona needs between l'Hospitalet, Esplugues, Sant Just and Sant Joan Despí if it wants to adequately project towards the 22nd century.

These are the stem cells with which we plan to fill the gap between l'Hospitalet, Barcelona, Sant Just, Esplugues and Sant Joan Despí, in an operation that had to rely on the intelligent mutual understanding of the Barça of Messi, Iniesta, Luís Enrique, Sergio Busquets, Suárez, Neymar and so many others, but ultimately the mutual understanding of a Barça capable of thinking about its future, and the future of everyone. But it wasn't possible.

A few years ago, in a bold decision by Pep Guardiola, Barcelona decided to definitely transfer training sessions and coaching staff to the Joan Gamper sports complex in Sant Joan Despí. The city of Baix de Llobregat has a strong and intelligent mayor; and depending on how you look at him, an old school mayor: Antoni Poveda. Poveda has a dream for his city and, in part, attracting the Barcelona first team to go to Sant Joan on a daily basis allowed him to obtain one of the first foundations for his dream: turning the motorway that crosses Baix Llobregat, from the Barcelona exit, into an urban avenue. Núria Marín, the mayor of l'Hospitalet, is always on the lookout for ways of intelligently using what Barcelona can offer without losing an iota of my home town's personality (I'm from l'Hospitalet, whatever my identity card says). She shared the need to definitively rethink the empty space behind Diagonal, between Esplugues and Sant Just, Barcelona and l'Hospitalet with me and Poveda. The mayor of Sant Just, Perpinyà, was another ally, as was the mayor of Esplugues. The regenerative vision, the vision of urban reconnection in those places that had been so heavily punished by the crisis and a lack of action, was a hopeful starting point: we were going to propose to Fútbol Club Barcelona building their new stadium in the focal point of these five municipalities, very close to their sports complex, in touching distance of the metro lines 2 and 9, and at the entrance to Barcelona through Diagonal, with a façade looking down the avenue and growth towards the Carretera de Collblanc. All of this only five hundred metres away from the current location of the stadium.

The operation made it possible to carry out another dream of urban improvement, already proposed in the study that the Metropolitan Area of Barcelona had simultaneously carried out: how to sort out the scalextric-like exit from Barcelona through Diagonal. The study, carried out by the prestigious architectural firm Batlle i Roig, anticipated and forecasted the configuration of Diagonal as a whole, up to the Llobregat river, as a long avenue. However, what was most striking about it was the transformation of that knot into a great plaza, with vehicles passing over it, just in front of what would have been the front of the new stadium, to Diagonal. That was where the construction of a hotel and space for offices and services was planned. Right next to it, on land in Esplugues that had been designated for more square metres of housing and businesses, the city would have been configured around the activity induced by the presence of sports, but also museums, centres of expertise, health centres and spaces for Barça to conduct business. Likewise, the construction of the Palau Blaugrana, between Avinguda de Manuel Azaña and Carretera de Collblanc, would have given the façade of the metropolitan city a new reason to establish residences and activity based on multilateral business. All at once, this space didn't seem to belong to anyone, places that we know of due to failed plans like Caufec,[16] or due to having the largest concentration of prostitution in Barcelona. They could have become the western centre that Barcelona so desperately needs, to bind the true metropolitan city together.

Until now we haven't talked about the university, which owns the land on which we proposed to build Barcelona's new facilities. The Universität de Barcelona, in the shape of their provost Ramírez and his team, were always on our side. I remember the day I proposed the idea to the provost, unveiling plans and identifying spaces. We spoke

---

[16] An urban mega-project started in 1991 in Esplugues; the project was plagued by corruption for the day it began.

about the urban campus, which ran from the northern side of Diagonal to Riera Blanca and Travessera de les Corts. A space flanked by the Parc Científic and the adjacent faculties, with student residencies and sports facilities that, incidentally, had been managed by FC Barcelona itself, as we kept saying in the conversations held with all the stakeholders. We had even agreed that the Masia, the former residence for future footballers where people from Guillermo Amor to Pep Guardiola had lived, would remain in the club's hands, as well as the pitch at the Camp Nou. We had begun to work on a project, thanks to virtual reality, that would memorialise the pitch's life through images that could be captured on Smartphones as you approached it. Barcelona's various universities would have seen their unique character reinforced in Europe and the world, due to forming part of one of the most integrated learning structures in any city. Not to mention the obvious benefits to the populous districts of les Corts that, when the space the Camp Nou had occupied was put to community use, would have seen the Jardins de la Maternitat grow to become one of the most important parks in the entire metropolis.

The decision taken by the Barcelona board was another. I remember how the head of the club's heritage told me that they had commissioned a study by a North American specialist in stadiums. He had told them that it wasn't possible to build it where we had proposed due to a slope in the terrain. This was in fact one of the best reasons to build it at that part of Diagonal, right at the beginning of what would have been the new road to the banks of the Llobregat. We did manage, however, to join Barcelona's cycle lanes to Baix Llobregat's at exactly the same point. It's not much is it . . .

In short, the new Camp Nou will stay where it is now, we will continue to see great goals there and I'm sure that it will also serve as a benchmark for the city, as it has done until now. The architectural project and the urban planning around it is powerful, and there's no doubt it will allow for a new interaction between the stadium's surroundings and the city. We'll have to wait and see.

Barcelona, who more than twenty five years ago had taken advantage of the pull of the Olympic Games to colonise and reenergise the western part of the city, could have followed its own route to the west. By not doing so, it didn't follow the path London took, by copying Barcelona and Manchester's models in its Olympic East End. The team was made up of Ricky Burdett and Richard Rogers, who are so influenced by Barcelona but at the same time have their roots in traditional British town planning of the thirties. They decided to push the city beyond Canary Wharf, combining audacity and a vision of the future. Canary Wharf had arguably been the most important developmental point in London as it moved eastwards. The Olympic operation was conceived with a much more balanced approach from the point of view of what the city should be, in the sense that it was thought of as a flexible platform between the social and the financial. The combination of sports facilities with affordable and high-quality housing, more space for businesses of all kinds and, above all, massive public transport connectivity, set the tone for this operation. The installation of important cultural facilities set the tone for the area. In the meantime, in Canary Wharf, London is building a public transport hub that will have affordable, as well as high-end housing, built on top of it.

During the 19th century, in the shadow of social utopias, the city of the industrial revolution was asked for basic services like water, gas to light the streets, electricity wherever it was at hand and steam to make trains and factories run; and, it's clear, housing and work, housing and work. During the 20th century we have asked the city for all of that, as well as efficient mobility as far as possible, with a correct but difficult balance between public and private transport; access to quality education and public health services for all; social care that will reach all the vulnerable layers of society, in addition to sustainability and respect for the environment towards the end of the century, already anticipating the environmental challenges of the 21st century after so much abuse. And of course: housing and work, housing and work.

HOW DO WE LIVE? WHAT TYPE OF CITY ARE WE HEADING TOWARDS?

The 21st century is the manifestation of inequality on a global, but also local, level. Our century has cracks in it: nothing seems to completely fit, and what's worse, now we have grown up and we know that there are no utopias that are going to save us or provide magical solutions to all of our problems. We know that those who talk about them, those who sanctify anti-globalisation, the market, or populism of any kind, are speaking half-truths, or openly lying. They take advantage of a prevailing lack of hope in order to prosper, using false messages that don't have good conclusions. Ours is the time of the drug cartels, the time of failed states. Of religion used as an excuse for all kinds of abhorrent deeds, forgetting the need that humans have for transcendence. But it's also the time of a better prepared and more conscious youth, the time of the technological revolution at the service of the people; the time of connectivity, where it's complicated to make something go by completely unnoticed. It's the time when India, China and so many other countries are emerging from the absolute poverty of millions. We are living in a simultaneously split and integrative time. A time of scarcity and overabundance. Yes, overabundance, because in many ways the new crisis that a large part of our society is going through is a crisis of excess.

The answer has been to show that there is a new form of citizenship, closer to expectations of self-fulfilment, closer to the transcendental. Some young people and many adults, rather than making sense of life, look to fill it with time for themselves, in a withdrawal that is quite possibly a product of the ennui provoked by the untrue utopias we have talked about, and abused. It is therefore a kind of post-post-modernism that has repercussions on how life is conceived, and as such, the way cities are conceived. Only in this way is it possible to give an explanation for these urban movements of the ephemeral that revolve around the occupancy and self-management of spaces. It's obvious that we need to understand what's coming next, so that the city can provide an adequate response. Cities, in addition to everything that has been asked for over the last hundred and fifty years, are now being asked for life experience, work, a decent home, a good stock of talent, global connectivity, to provide the information technologies that are available to all, and infrastructural connectivity that allows us to enter and leave them easily. Cities are asked for opportunities for all, equality, social integration, quality of life, security . . .

No more than twenty or thirty years ago, the most educated youth would have given almost all of the little they had to be accepted to one of those great auditing or consulting firms that subjected them to a sort of servitude whereby they would work sixteen hour days, with the promise, what's more it was true, of one day becoming a manager or even a partner of the firm if they were able to hold out. Respect, a good salary, a good exit on the other side of their professional careers. Not to mention the architectural firms of the time, where young professionals crawled between drafting tables while the star architect received prizes and commissions. But in the end it allowed you to build your own office from which to mount your success. The country was good for that and a lot more. Today that way of living the present with the aim of nurturing the future is rapidly disappearing, people are no longer content with promises in the future. Today people want to live the present with intensity because it's all they have, and it's within this framework that the concept of experience is revealed to be fundamental. Hence, the city must seek out that experiential direction, that on the other hand is intrinsic to it. If cities like Barcelona, London, New York or Sydney triumph, it's because the balance between life experience and global training is remarkably efficient. Furthermore, if Barcelona is ahead of the world's other cities, it is down to its blatant social choice, for trying not to leave anyone behind, for valuing the public sector in a way that is not at all evident in the aforementioned cities. And for having adopted the more complex option some time ago.

Understanding how we live and where we are going is key to designing the city:

1. A new way of living where subscription is more important than possession. What else apart from Spotify or Netflix? Few people still buy CDs. The majority of people buy music as if it were Pic 'n' mix, renting access and fleeing from physically owning it. Think about the progressive *uberisation* of our streets. Or the Airbnb phenomenon. So many years of talking about cooperating without middlemen, when technology implants mass ground-breaking models, we logically enter into crisis. But these new styles of life are here to stay, because they offer more and better services that are more efficient than those that have sprung from industrial society and the first service society. It's the time of inbound marketing, the time of influential figures on the internet, of boredom of direct advertising and of the massive growth of the recently-coined "below the line" advertising through the internet of bloggers. The city, as a body, must think about how to position itself in the world and contemplate these new ways of living, but also how to attract global attention thinking about the new ways of generating image. All of this will have a physical and morphological translation, in terms of

building the city and sewing together the bits and pieces of cities that have been forgotten about during the great crises of the last third of the 20th and start of the 21st centuries.

2. Vibrant spaces, in which generative exchange is imposed on privacy. Today it's no longer possible to talk about modernity in the design of public and private space without taking into account co-working, without making offices spaces for exchange; without giving squares or streets that key point of humanity that makes them comfortable. It's the time of 3D printing, the time in which we are beginning to understand that we must return production to the heart of our cities, the time of Manufacturing Associations, those revolutionary civic centres where we can make anything from cups to houses, but ultimately it's the time of interacting through manufacturing, making, assembling, creating. It's the time of citizens who have decided to take control of what they publish. The time of advanced logistics, the disappearance of queues through the advanced management of flows of people. All of this contributes to creating spaces that are more human and rational, more warm and efficient.
3. The social transformation that we are living is loaded with authenticity, and by extension, identity. I'm referring to the authenticity of what we build, which needs to be vernacular, our own, human, contextualised within the culture in which it is introduced. It's the time in which we can finally understand the depth and truth contained in Isaiah Berlin's famous essay *The Hedgehog and the Fox*. In this essay on Tolstoy's view of history, Berlin foresaw this new way of defining modern professionals, experts in one field and amateurs in many other complementary fields. Using Berlin's example, as well as the Greek poet Archilochus, let's remember the verse where he says, "A fox knows many things, but a hedgehog one important thing." Cities are faced with the challenge of maximising this world without losing opportunities with urban monocultures, closing spaces, obstructing unplanned creativity. Authenticity is not banal. Authenticity is not easily franchised. Today authenticity is both artisanal and digital at the same time, in such a way that serialisation is made possible by combining it with personalisation, something that was impossible

in industrial and post-industrial times dominated by Fordist systems of production.

In short, the stem cells of the new city that I spoke of above are made of new ways of living, new, vibrant and identity-filled spaces. These are the three starting conditions for the city we must learn to build in order to respond to the challenges that will successfully launch us into the 22nd century. All of this bursting, of course, with equity, equality of opportunity and redistributive capacity.

Note that the key to what must happen always revolves around the density of private, civic and public activity, the combination of affordable housing and free pricing, the possibility of working near where you live, high-quality public space for shared leisure (culture, walks, nature), hyper-connectivity and public transport. That is to say, housing, work, leisure and transport. This is the city in which all types of human activity are intensified, are condensed.

The recipe for your CONDENSECITY:

- Affordable and available housing
- Co-working spaces and active commerce at a ground floor level
- Markets and businesses that are close to the community
- Open culture and a few ounces of elitism
- Sport and leisure as a social good
- The right to walks, views and water as urban elements
- Hyper-connectivity

This is something I have sometimes called the *Condensecities* model, that place where a creative dialogue takes place between the diverse uses of a city. In Barcelona, for example, Eixample's success is due to the balance between housing and activity, planned space, and space that's open to interpretation. We're the antithesis of London's fractality, for example. But we're equally close to the freedom that can only be guided by the rhythm of the orthogonal grid. On this chess board anything goes, and that "anything" works as long as it responds to the criterion for condensation of activity, possibly inherited from the old Chinese district or the Rivera neighbourhood, overused spaces in which Barcelona, despite everything, continues to be recognised.

THE SAGRADA FAMILIA: CITY, TOURISM AND INTELLIGENT URBAN PLANNING

It's precisely in another of Barcelona's essential locations, Eixample, where the city is rolling the dice. I'm talking about the Sagrada Familia and its surroundings. Many might think that everything started with the time-honoured Antoni Gaudí and his brilliant madness: to build an expiatory temple, a cathedral of the people and for the people, in the heart of Barcelona that Engels loved for being the most bombed and most revolutionary in Spain. Engels knew very little about Barcelona, it's clear, and clutched at straws. Gaudí, on the other hand, forefather of the maxim coined by the poet Foix, "I am exalted by the new and in love with the old," raised the spires of the Sagrada Familia in the centre of the newborn city. He did so on the condition that the temple would serve to atone for the sins of Christian parishioners. A true Catholic landmark, in other words universal, to give character to the new city. The condition meant that the new cathedral could only be constructed by way of donations – from the end of the 19th century until well into the last quarter of the 20th century. But Gaudí was going to mess it up, as the prince of 20th century Catalan poets, Josep Carner, predicted, at the time a Noucentista, when he wrote about the great architect, the high priest of a very particular style of Modernism:

Tothom n'ha sentides dir
d'aquest gran senyor Gaudí
que cada hora – no s'hi val! –
fa una cosa genial
i no deixa viure en pau
l'home savi ni el babau.[17]

The idea of building an expiatory temple dedicated to the Holy Family on new land in Ensanche belonged to the bookseller Josep Maria Bocabella, inspired by the priest Josep Manyanet (canonised in 2004), the founder of the religious orders The Congregation of the

[17] Everyone has heard, of this great man Gaudí, who every hour (it's not fair!), does something great, but he lets neither wise men, nor fools, live in peace.

Sons of the Holy Family and The Missionary Daughters of the Holy Family of Nazareth, charged with promoting the worship of the Holy Family and Christian education for children and young people. In order to raise funds, Bocabella founded the Association of Devotees of St. Joseph, and for the construction of the temple he acquired an entire block of land in Ensanche in a place known as El Poblet, near the Camp de l'Arpa in Sant Martí de Provençals, between Carrers de Provença, Mallorca, Marina and Sardenya. The plot cost 172,000 pesetas at the time. The project was first entrusted to the architect Francisco de Paula del Villar y Lozano, who devised a Neo-Gothic design and dismissed Bocabella's idea of replicating the Santuario della Santa Casa di Loreto (Ancona, Italy), which is supposed to contain what was the house of Mary and Joseph in Nazareth. Villar's project consisted of a church with three naves, with typical Gothic elements like lancet windows, exterior buttresses and a tall needle-like spire.

The first stone was laid on the 19th of March 1882 (St. Joseph's Day), in the presence of the then Bishop of Barcelona José María Urquinaona. Gaudí attended the ceremony, since he had worked as Villar's assistant on several projects; at the time he could not have imagined that he would become the project's architect. The works didn't start until the 25th of August 1883, being awarded to the contractor Macari Planella i Roura. In 1883 Villar resigned due to disagreements with Joan Martorell, architectural advisor to Bocabella. The projects was offered to Martorell himself, but upon refusing it they turned to a young, thirty-one-year-old Gaudí. He had been an assistant to Martorell for several constructions, a fact that triggered the recommendation of the recently licensed architect, who still hadn't carried out large works. Taking charge of the project, Gaudí modified it entirely, apart from the already built crypt, and he imprinted his peculiar style on it. During the remaining forty-three years of his life he worked intensively on the project, the last fifteen exclusively. Such intense dedication has an explanation, aside from the magnitude of the project: Gaudí defined many of its aspects as construction progressed, without having previously specified them in his plans and instructions. Hence his personal presence in the work was extremely important.

From 1985 onwards the project was managed by the Construction Board of the Expiatory Temple of La Sagrada Familia, an ecclesiastical foundation created to advance the basilica's construction through donations and public and private initiatives. Its current president is the Archbishop of Barcelona, Juan José Omella, and the former was Cardinal Martínez Sistach, a protagonist of part of this story. We'll come back to him.

Gaudí estimated that the construction would last centuries. For this reason he proposed to the Construction Board raising and finishing the Birth Façade so that the generation that had begun the works would see something finished (an early victory!) and, simultaneously, the finished façade would serve as a stimulus for future generations to finish the temple. His proposal was accepted. In 1923, still during Gaudí's lifetime, the calculations for the structure of the naves were published in the Association of Architect's bulletin, signed by his assistant Domènec Sugrañes. Those calculations have been used to continue construction, although they have been adapted in part to comply with current legislation. Gaudí, aware that the construction of the temple would be carried out by later generations, tried to define the project through blueprints, but, knowing that he wouldn't have enough time during his life, made three detailed, plaster models at a scale of 1:10 and 1:25 of the most significant parts, with the hope that they would be used as examples for the rest of the building. Gaudí designed with three three-dimensional models the central nave, the sacristy and the Glory Façade. The model of the principal nave would serve as the blueprint for the others naves and the model of the sacristy for the central spires.

In Gaudí's lifetime only the Birth Façade was built, with sculpture by Carles Mani, Llorenç Matamala and Joan Matamala, relying on drawings by Ricard Opisso. The architect only saw one of the spires crowned before his death, the St. Barnabas spire. Upon his death, his assistant Domènec Sugrañes took over the work (between the years 1926–36), who finished the three spires that remained on the Birth Façade.

On the 20th of July 1936, two days after the fascist military coup that started the Spanish Civil War, groups of anticlerical savages set fire to the crypt, the workshop where Gaudí had worked that contained his sketches, miniatures and models, which were almost completely destroyed. A few days after the destruction the architect Lluís Bonet i Garí requested the rescue of the broken fragments of models and thanks to this intervention a municipal brigade recovered the fragments, and they were stored. With these items, and others that were buried and recovered later on, and the photographs that were conserved of the original models, from 1940 onwards, Francesc Quintana, Isidre Puig i Boada y Lluís Bonet i Garí restored and rebuilt the models, drew up their plans and built a new replica of the model of the main nave on a 1:10 scale, which can be seen today in the basilica's museum.

When the construction of the Sagrada Familia was resumed in 1944, it was first necessary to define how to continue the temple's construction whilst being as faithful as possible to Gaudí's ideas. At the fore of this gigantic task were the architects Francesc Quintana, Isidre Puig i Boada y Lluís Bonet i Garí, whilst Jaume Busquets took charge of the sculptural work. Subsequently, when the Passion Façade was built, the main ensemble of sculptural figures was entrusted to Josep Maria Subirachs. The latter's works generated certain controversy due to the fact he created totally contemporary sculptures that were far removed from the Realist style Gaudí had employed on the Birth Façade. Likewise, the Japanese sculptor Etsuro Sotoo collaborated on some of the sculptures of the Birth Façade. From 1987 until 2012 the works were lead by Jordi Bonet i Armengol, at which point he was replaced by Jordi Faulí i Oller.

Let's look at the Gordian knot of the Sagrada Familia story: when the works began it was located on disused land, but it was soon integrated into the city's rapid development at the start of the 20th century. In 1905 Gaudí carried out a project that would include the Sagrada Familia in the Jaussely Plan, the new project designed to connect Barcelona's expansion district created under the Cerdà Plan with the new municipalities that had been added: he envisioned situating the temple inside an star-shaped octagonal garden area, which would have provided an optimal view of the temple from all surrounding areas. Finally, due to the cost of land, he scaled back the project and settled on a four-pointed star that allow a wide view from all its corners. However, Gaudí's plan was never implemented: in 1975 Barcelona City Council carried out an urbanistic study that envisaged the creation of an area around the Sagrada Familia in the form of a cross, with four squares of landscaped gardens on each side of the temple. However, currently only two of these squares exist, and creation of the other two would involve the demolition of several buildings, the majority of which had been consolidated after the approval of 1976's General Metropolitan Plan. The ideal solution to frame the Sagrada

Familia in the environment it deserves does not exist. This was the premise of the job that I entrusted to the Council's Project Management Team, on the basis of five conditions:

1. Access to the Glory Façade should maintain a sense of spectacle, immediately ruling out the "Via della Conciliazione"[18] effect proposed in the Metropolitan General Plan, bringing us close to the Borgian notion (the footprint Catalonia left on Roman city planning should be much more widely known and studied) of the entrance to St. Peter's Square. It was the easiest way to recover the effect Gaudí sought in his original plan.
2. Minimise the impact on existing homes, limiting it to the strictly necessary to be able to finish the Glory Façade.
3. Ensure the Glory Façade's connection to Carrer d'Aragó, and from there with the new configuration of the Plaça de les Glòries Catalanes, which would allow reinforcement of the new Parc Güell, Sant Pau, Sagrada Familia, Glòries axis.
4. Incorporate a plan for public facilities to the structural modification of the temple's surroundings.
5. Incorporate the impact of the mass influx of visitors in the temple.

These conditions were debated in several discreet meetings with neighbourhood representatives who, in private, considered them good. What was funny was that during this process we discovered that Glòries' neighbourhood representative was the same person as the Sagrada Familia's, who was the same person as Plaça d'Espanya's. In short, it was with him that I had to tone down Gaudí's ideas, of course, whilst he begged me to make sure no-one knew that we were meeting each other, given that the elections were coming up. I never used his word, nor the agreements we came to, when he and his people backed out of them.

In December 2013 the Council published a report containing several proposals for the urbanisation of the temple's surroundings, produced by the firm Estudi Massip-Bosch Arquitectes, in which they offered eight possible solutions: leaving it as it is; creating a 60-

[18] The road that connects Saint Peter's Square to the Castel Sant'Angelo in Rome.

metre-wide avenue to Diagonal, that would partially affect two blocks of buildings; creating the same avenue but thinner; creating a thin avenue to Carrer de València that would only affect one block; creating a wider avenue to Diagonal, completely demolishing the two blocks; completely demolishing the first block and creating a plaza similar to those adjacent to the Birth and Passion Façades; the four-pointed star sketched by Gaudí; and finally, a smaller variant of the latter.

Let's go back to the city that people live in. For many years my grandmother and aunt lived at the intersection of Consell de Cent with Sicília, very close to the Sagrada Familia. My brother Joan and I, when we were little, went to our grandmother's house to stay the night, or to spend the afternoon, from distant Hospitalet. I remember how our grandparents would take us to skate in the gardens adjacent to the temple, almost always empty and soulless. My grandmother said that when she was a girl they used to take her to see the church grow from Sant Boi, where she was born, and that perhaps we wouldn't see it finished either. Joan and I skated alone all afternoon, whilst the workmen built the temple, wearily, at the variable rate of donations. It was the age of splendour of the city's trendy rich, who allowed themselves to claim that the best thing that could be done with the Sagrada Familia was to knock it down. My paternal family, PSUC[19] activists, Catalan and catholic nationalists, victims of reprisals, working class and self-taught, a mixture that's so typical of our country's people, never understood the amount of contempt put forward by rejectionists for something that everyone loved.

Everything changed in 1984, and it was nothing to do with the echoes of the '82 World Cup, nor with Orwell, nor with anything related to the '92 Olympic Games that were still so far away. The revolution arrived with whisky, and not exactly from the Scottish Highlands, but from Japan. Everything started with an advert on television for Suntory whisky, one of the best in the world, as demonstrated by the awards it has received. In Japan Suntory adverts are what the Freixenet bubbles were for a long time in Spain. The legend goes that the Suntory team were going to Scotland to shoot the advert but a storm meant the plane had to be diverted to Barcelona.

---

[19] Partit Socialista Unificat de Catalunya, The Unified Socialist Party of Catalonia.

The team had to spend a few days in Barcelona, during which the film crew's managers spent their time seeing the sights of a pre-'92 Barcelona that was not at all touristic. It was love at first sight: the Suntory team found something far better than the Scottish Highlands. They found Gaudí. And they decided to take him to Japan, in the form of an advertising campaign. Later on I will talk about the importance of correct branding to attract investment and talent. Barcelona, without knowing it, was playing the leading role in one of the most important image rebrandings in the world in recent decades. The advert's effect, which begins with the image of a flamenco dancer leaning on a Gaudí column with the words, "Barcelona, Spain, Antonio Gaudí . . ." was staggering. The arrival of Japanese investment in the automobile industry was joined by the mass arrival of Japanese people who, inspired by the whisky, threw themselves at the opportunity to get to know and to make public the Catalan architect. From that moment on, with the explosion of the Games and the subsequent explosion of tourism to Barcelona, the Sagrada Familia started to increase the rate of ticket sales to see the temple. Tickets that weren't sold as such, but as donations for the atonement of sins and to contribute to the construction of more edifice. So much money has been raised that today it is public and well-known that the Sagrada Familia will be finished in 2026, much earlier than anticipated by Gaudí, and by my grandmother.

This Japanese-rooted inflation has had collateral consequences. Today it's difficult for children of the age my brother and I were when we used to skate in the park, to be able to enjoy the tranquility of the little green we have in the centre of Eixample. The touristic masses invade pavements and avenues, shops that have been there for a lifetime are replaced by souvenir shops, in some buildings there are more flats for tourists than flats for residents. The heavy traffic created by coaches, the queues on the pavement, the crowds at the entrances to the metro, make everything more complicated. I spoke about it several times with Cardinal Sistach and the members of his curia. My feelings towards the temple and what it represents are more than positive. However, it's evident that the ecclesiastical hierarchy, which has gained so much from Gaudí, and from Japanese whisky, has to generously accept its success and socialise it. Specifically, I asked to speak about actions they could take:

1. To open the temple up to neighbours for a few hours a day so that they can pray, as well as opening up some spaces for their civic use, with the goal of reincorporating the temple into neighbourhood life.
2. To interiorise the queues of tourists, in other words, to create interior space to decongest the pavements and closest entrances to the metro, as a precursor to the definition of a more ambitious reception project underneath the adjacent plazas.
3. To finance the construction of an underground stop for coaches, a reception hall with a digital welcoming exhibition and a space for exiting visitors, by raising the price of entrance by one euro. Once the work has been financed, the tourist's euro will go towards maintaining the adjacent civic centres.

None of the proposals went any further. I humbly think it was a missed opportunity. We dedicated ourselves to studying the crowds through digital technology, we monitored them precisely, hour by hour, day by day, we monitored how people moved around the temple. It was one of the first instances of the use of digital technology for the redefinition of public space.

We were able to start working with the most active neighbourhood representatives on the redefinition of pavements, terraces, location of shops, mobility and general traffic. It was the prelude to choosing the definitive solution to the urbanistic surroundings of the temple. Cardinal Martínez Sistach withdrew from the process. Perhaps he was thinking about retirement, or about some Roman conclave. He certainly didn't think about the opportunity that was presented to him to conclude Pope Benedict XVI's visit, in which the Sagrada Familia shone with the light that the genius Gaudí had given it, the artificial illumination and those perfumed drops of malt, brought from the land of the rising sun.

Today, Suntory continues to break the mould in terms of advertising and generation of brand awareness. They do it like they've always done it, by looking for the most innovative idea, which doesn't necessarily mean the newest. Innovation is the art of finding different approaches to the seemingly obvious. There's something poetic about innovation. Innovation is like a Bach fugue. To innovate is to remove the veil that hides what was already under it. And to be able to innovate you need to be with the best. Suntory know this and turned the most cutting-edge innovation into its new way of presenting itself to the world. Suntory has united with 3D printing and begun distributing printers among its outlets. Why? For something so simple as personalising the ice cubes with which we'll brighten up our nip of Japanese whisky. Imagine Bill Murray in the advert in *Lost in Translation*, with freshly shaped ice cubes designed to look like the image he, or Scarlett Johansson, has sent from his laptop or smartphone. Of course! In 2005, when the film was shot, that wasn't possible. Today it is and Suntory, through their campaigns, has become the brand with the most advertising hits in history: in two weeks more than five million tweets from forty-seven different countries; in two weeks more than two hundred establishments voluntarily asked if they could install their ice cube machine; the media coverage was spectacular; 346 million views . . . Winner of the Cannes Lions Festival of Creativity. Have a look at

> http://br.adforum.com/award-organization/6650183/showcase/2014/ad/34499102,

and enjoy. Afterwards, think about whether you've surrounded yourself with the best people, whether you're really creating the future, whether you're putting it within reach of your city or just enjoying the glass of whisky that someone else has made, with its same-old, boring little ice cube that observes you sadly as it melts, just as you do.

FROM PITO TO DIDO

The lesson to be taken from some of my failures in Barcelona perhaps has to do with my excess of passion, audacity, ambition and true love for everything innovative, for the best and most risky, in favour of society, in favour of rearming it to improve quality of life, to give it more opportunities, to minimise inequality without verbosity, to give power and a medium to the ability to work and to those who believe in individual effort as the basis of all prosperity. It's advisable to see the world and think a lot. To see the world and imagine what we want to be in fifty years, knowing that perhaps we won't see it, but we have done everything possible to make it happen.

Hence it is advisable to go, for example, to MIT (Massachusetts Institute of Technology) and think three, or four generations in advance. When you enter MIT's Center for Bits and Atoms directed by Neil Gershenfeld, it's as if you were entering Barcelona's Servei Estació[20] warehouses thirty years ago: corners full of toolboxes, pipes opened lengthways through which fibre-optic beams pass, 3D printers combined with laser shuttles, water jet cutters that slice pieces of metal four inches thick, computer screens, transparent keyboards and crystals filled with mathematical algorithms. I ask the Centre's director, who invented the Internet 0 concept, the Internet of Things, what kind of scientists choose to join his team, "We're the misfits," he tells me, smiling. At this moment I remember Clark Gable and Marilyn Monroe acting in John Ford's film, with the script by Arthur Miller. I also remember the first advert of Steve Jobs' second stint at Apple. The director carries on talking, "We're the ones who don't fit, who live outside the rules . . . " Neil tells me with a mischievous expression. If I had to describe him I would refer you to any photo of Albert Einstein with a 21st century twist: white shirt, Dockers and white trainers . . .

At Columbia University I see the Guastavino files, in dean Mark Wighley's hand. A part of New York history is there, for example Grand Central Station, or the old and missed Penn Station. To go from the headquarters of the School of Architecture you have to pass through the university's library, which is the heart of the main building. A good metaphor for your city. Mark tells me that what's great about Columbia is that it's a university installed inside a library. Another great metaphor! Above it, the classrooms, student residencies, lecturers' offices. Noise, slides, seminars, cubicles for discussion in pairs, spaces for reflective dialogue and critical thinking. The thing Columbia's students most enjoy is testing their teachers, confronting them with their contradictions. Below, in the library and basement, you can find the archives of the man who extended the Catalan vault through North America (you have to visit the Oyster Bar in Grand Central Station, New York, study it through Catalan eyes and discover the power of the genuinely local). Order, reflection, silent study, the foundations of a society.

Harvard breathes power through every pore. It's a university that has been designed to inject character into those who pass through its classrooms. Harvard is the fundamental structure, it's study dedicated

---

20 A business that sells building and design materials from its warehouses.

to strengthening institutions. When a doctor gets in to the School of Public Health, like my friend Belén Fraile, a young *madrileña* with a Fulbright Scholarship, they are inoculated with the essence of profoundness, with the fragrance that standing adds to personal standing. The same could be said for anyone who passes through any other faculty, let's say the School of Architecture. Belén's husband, Dani Ibáñez, ex-student of the Institute for Advanced Architecture of Catalonia, was asked to be part of Harvard thanks to the work he carried out in such an informal and independent place as the IAAC (Institute for Advanced Architecture of Catalonia). For Dani and Belén, Harvard has represented opening themselves up to understand the deep functioning systems of government. Opening themselves up to understand critical thinking, often doubting what they are receiving as teaching, considering it merely a starting point.

After mulling it over more than a few times I've arrived at the conclusion that only societies that value their misfits, societies that empower them, finance them, that turn their Zuckerbergs into heroes, their Gershenfelds into examples, their Mercès and their Beléns into models to follow, move forwards. Are we brave enough to do it? We have to break the systems that stagnate and with narrow-minded egalitarian arguments forget that only through merit, appreciation of intelligence and the ability to sacrifice, do societies advance. We have to make sure that the best people, wherever they are, are given the opportunity to prove they're the best. We must implement models that promote excellence, without leaving anyone out. And this goes for those who don't follow, of course. But above all for those who are ahead of the game.

After the seminar we sat down with Neil and started to take copious notes. We talked about the fact that today cities follow the "PITO" model: "Products In – Trash Out." Neil suggests that the new industrial revolution consists of making cities DIDO: "Data In – Data Out." We discussed it with him and his colleagues and began the steps to turn it into a research project in which the Universität Pompeu Fabra's I2CAT, lead by Dr Sallent, was implicated from the first moment. We'll need engineers, architects, lawyers, doctors, citizens, businesses . . . But above all we will need a change in general mentality, a change to the way we conceive our society, if we don't want to resign ourselves to seeing other people think of new ideas and do it all. And what's good for my country, Catalonia, is good for yours, my friend, wherever you may be. That's the well-executed Smart strategy designed for future generations. However, it's very difficult to talk about three generations in the future without having a good plan for the immediate future.

THE EARLY VICTORIES PLAN. MICRO-URBANISATIONS

Citizens are patient and know how to wait until works are finished, until a plan is implemented. Citizens however, are fed up with promises that aren't fulfilled. And in this sense, the long term, plans developed by the cleverest scientists or the most expert planners can create mistrust and boredom. So we thought it was very important to equip ourselves with a plan that would drive us to align the long term with the short. This was how Barcelona's micro-urbanisation plan came about. On the other hand, we wanted to demonstrate that we were capable of departing from the two-year paradigm, that which seems to force us to think that nothing is possible in our city in less than two years.

It involves city planning at a level close to citizens, that allows for small, simple and flexible actions that improve people's lives. To develop this idea I was inspired by the conference Federico García Lorca gave on the 17th of October, 1926 on the Granadian poet Pedro Soto de Rojas. In the lecture García Lorca affirmed that the city's authentic aesthetic is the love of the small, the delicate, the intimate. The micro-urbanisations programme looked to convert empty or deteriorated lots that didn't have a concrete short-term use, into comfortable places, adaptable for residents' use and for different purposes, cohabitation, living space or work. The objective was, therefore, to convert these lots into spaces of opportunity for urban revitalisation, social dynamisation and, in many cases, the renaturation of the city. The chosen projects prioritised the potential of social dynamisation, the implication of the environment, the reversibility of actions, a reduced ecological footprint, excellence in design, and control of costs and economic viability. The sites were decided district by district, taking into account residents' requests and insight into the needs of each neighbourhood. Among the micro-urbanisations implemented, a large variety of programmes involved the modification of sports facilities, shaded area for relaxation, playgrounds, green areas, spaces that connect . . .

Micro-urbanisation on the Germanetes block, in Barcelona's Eixample district, combined with the Buits Plan, the programme of social and urban revival.

Micro-urbanisations

http://habitaturba.bcn.cat/blog/es/programa-de-microurbanitzacions-urbanisme-de-proxi-mitat/#.WGeDepIeaRs,

which have since been exported to the whole world, are low-cost interventions (between £53,000 and £90,000) with high civic impact. As I was saying, it involves recovering public land, pending allocation or in the process of urbanisation, on which temporary housing developments are created with the help of architects and city planners of all ages.

These early victories helped to dignify many spaces that were in very precarious situations, if not complete abandoned. In micro-urbanisations we discovered ways of acting quickly, alternative uses for materials and disused urban furniture, spontaneous management of green urban spaces, use of communication technologies, practical management of creativity, and we even managed to better understand the role of provisionality in the city. We came to the conclusion that often nothing is more permanent in a city than that which has been designed to be provisional.

Micro-urbanisations, like the later Buits Plan or plan for the development of empty urban spaces

(http://ajuntament.barcelona.cat/ecologiaurbana/es/plan-buits),

which was nothing more than the transfer of disused spaces to civic entities so they could utilise them for a few years, has become a blueprint for action in places as far-flung as Latin America and India. It is precisely in India, in the heart of the transformation processes of the one hundred Smart cities launched by the government from Delhi, where micro-urbanisations are becoming strategically important, given that they offer the possibility to see and to verify what Indian cities aspire to in terms of public space. There, micro-urbanisations are called "place-making" actions, and Pune, under the leadership of the great commissioner Mr Kunal Kumar, was the city that launched the process. The first micro-urbanisations acquired a educative quality that is even more important than in Barcelona, with more profound intensions, to create a shared identity through urban action, with a view to deploying the Smart strategy, as well as the need to rely on citizens' involvement. This can be seen in the basic scheme for the reconversion of Pune's *nalas*, or open sewers.

It takes great imagination to arrive in Pune and see the *nalas* as a great opportunity to, using them as the crux, generate the change that the city needs to experience in the coming years: through its *nalas* Pune will overcome its unbearable traffic, the pollution that contaminates everything; through its *nalas* Pune is building a new future for everyone that enjoys the creative liveliness of one of the cities with the most potential in Asia. In Pune there is no fear of mixing meaningful intervention in the present with a vision of the future in three generation's time, given that one cannot exist without the other.

10

Energy Through the Internet

On the 1st April 2015, at 5:49 p.m., Alan Rusbridger, editor-in-chief of *The Guardian*, contacted people across the world that formed part of the newspaper's support group to, "Let us know he had some big news for us." Rusbridger told us that Guardian Media Group would start the process of disinvestment of their more than £800 million portfolios of investments in fossil fuels that very day. It was the largest global fund that had ever been subjected to a disinvestment on this scale.

The news was very important because it represented a shift away from symbolism and started a process that wasn't neutral for a media group as important and respected as *The Guardian*. No-one fails to pick up on the relationships between advertising revenue, the financial industry and the media world. Even more so today, when the press finds itself immersed in the very difficult reconversion process that everyone talks about but no-one really knows how it will turn out.

With this exploit, *The Guardian* helped to give an international boost to those of us that defend cities' energy self-sufficiency as the only way to make global change possible. I'm referring to a change in geostrategic relations that has to allow the Western world to impose itself on the dangers of all types of totalitarianism, including the ones we have created ourselves. "The Western world" has to be understood as the world of representative democracies, the world of individual freedom, the world of social cohesion, of equality of opportunity, rule of law, legal security, freedom of enterprise, right to private property . . .

It's the option that opposes the totalitarian populism of the right and left. To carry out the necessary change a radical movement lead by cities needs to take place, a movement in the same sense that *The Guardian* proposes and that I propose in this essay: energy self-sufficiency, distributed production, zero emissions. We know it's possible and we have to start moving decisively in order to achieve it. We can't carry on financing extremely expensive military operations, financial structures and the export of pain (inequality) to maintain our standards of life. It's been some time since Thomas Friedman illustrated the direct relationship between the price of oil and the power of oil and gas dictatorships to determine the life of the world. How much injustice have we financed by interchanging well-being with a greater oppressive capability? The game is clear: either we start distancing ourselves from fossil fuels, or it will carry on being impossible to base our liberal democracies (the only ones that guarantee freedom, equality and achieving well-being) on ethics. We cannot defend ethical principle in one place and make it seem like they don't count in other parts of the world.

Let's return to Barcelona. We already know today that by turning 30% of the city's roofs into solar panels and adding a substantial programme of wind-powered generators to the city's windier neighbourhoods, Barcelona can be self-sufficient in less than thirty years. The change is clear and is based on some deeply ideological motives. Self-sufficiency underpins freedom, creates work, requires large investments and therefore fuels the market economy, which we know is most efficient in the distribution of resources, goods and services.

Alan Rusbridger explained to us that the step his group took would not have been possible without the support of many people who, like all of us, defend the idea that it's not possible to persist funding searches for the supply of fossil fuels. He asked us to participate in the campaign

#keepitintheground

to keep pushing to change the current way we produce energy. He also asked us to explain the idea to as many people as we could . . . The logic behind the campaign is sound. But it would be even better if it were the cities of the world that discovered that self-sufficiency is a practical, possible and genuine method of bringing change to fruition. The road to a better world, with more opportunity for everyone.

The Western World fluctuates between two problems: between scarcity and frenzy. The shortage of energy and the gold rush . . . it's simple. Both problems are old acquaintances in Catalonia, which brought about an industrial revolution with a serious energy shortage and suffered a gold rush that took the country to the limit, little more than a century ago. It's a good idea to read the great writer Narcís Oller. The Catalan industrial revolution took place without first-class energy sources. It was the European exception. It's not that we didn't try to harness the rather scarce supply of water in our rivers, but it soon became obvious that it wasn't going to work. Catalonia revolutionised industry without British, French or German coal. Northern Europe sought the conquest of colonial empires, combining the power of energy sources with technological advances. We could only rely on technology.

Travelling on the London Underground, with its characteristically Edwardian mien, is a lesson in history. The colonies fuelled the markets and laid the foundation for a world order that somehow lasted until the end of the Second World War. Meanwhile in Catalonia, we lived by a combined regime: on the one hand learning from those who knew best how to compete in markets that were favourable to us. On the other, trying to use political influence in Spain to preserve an industrial base to help it resist hard times and its rich cousins north of Paris. The remnants of the colonial empire gave what they could. Coal however lost its importance and dominance shifted towards oil. The geostrategy of the second half of the 20th century came into play and exploded after 1945 with the collapse of the British Empire and the birth of the Arab states, a principal source of energy resources and global problems. The United States had never been better off, it seemed to reach its pivotal moment with the collapse of the Soviet Union. It was the end of the story . . . and the start of the global gold rush.

Just like the protagonist of the colossal and now classic novel *Conversation in the Cathedral*, by Vargas Llosa, we no longer remember the day that the West fouled up; the day we decided to export the production things that were fundamental to us, somewhere else. We forgot to keep fighting for our industrial base and succumbed to the madness of cheap bulk, of reinforced concrete sloshed all over the place, and the madness imagining that someone would pay for everything in the end. Like when Gil Foix, the protagonist of the novel *La febre d'or* [Gold Rush], quits his job as a carpenter and decides to earn his living in the market on Barcelona's Career d'Avinyó. A glut of credit, a glut of brick, social and demographic changes that were impossible to manage, an excess of speculative deals that were vulnerable to the 21st century's phylloxerae. And in the meantime, what? Giant countries like China, Brazil, India, have become akin to empires that knock on the door of progress by mastering what had been the basis of European dominance: energy sources, financial resources and industrial capacity. All of a sudden, Europe and the Western world as a whole are finding the limits to their own model.

TOWARDS ENERGY SELF-SUFFICIENCY: CITIES ARE THE SOLUTION

We need to be brave and intelligent to create a new energy and industrial paradigm that doesn't make us retrogress to the extent that the principles and values on which we have built our social model (freedom, equality, democracy, basic human rights) collapse. The city, your city, our city is where to resolve this issue. It is time for a new energy paradigm based on renewable sources that will rapidly disconnect us from the oil loop; a paradigm that creates new methods of industrial development in accordance with the sustainable and ecological ethics, already the majority in the West. Sustainability and progress should not be seen as resources to be harboured to excess, but as methods to guarantee well-being, culture, wealth to as many people as possible. In Barcelona we started the journey towards self-sufficiency in 2011. In Copenhagen, for example, they hope to be CO_2 neutral by 2025. We must not leave our cities on the peripheries of this new paradigm, we are fortunate enough to be prepared to become pioneers and lead the way. We've endured all the rushes and fevers: gold, brick, forgetting the foundations on which we had built progress.

It's a good moment to reflect on cities' possibilities with energy. I'm referring to the energy that we use to illuminate our streets and homes, to power factories and engines, to fill up the batteries in electric vehicles or to run our domestic appliances. Political independence should be linked to a plan for energy independence, which in turn should involve a commitment regarding the distribution of production. It's nothing more than applying the principles of the internet to energy production. We can't just rely on large power plants, or try and guarantee that far-flung energy sources will provide us with sufficient power for the country's hardware, we need to decide how we are going to disconnect ourselves from a world that is inclined to become a generator of uncertainty, contamination and even pain.

The production of energy in a distributed way is not a chimera. Imagine that you have an electric vehicle in the garage underneath your house, if you're lucky enough to live in a building with its own parking. Imagine that you charge this electric vehicle's battery through a combination of photovoltaic panels and wind turbines installed on your house's roof. Imagine that you only use the vehicle at weekends and therefore during the week, first thing in the morning and in the evenings, the energy that has been accumulating throughout the day in the car's battery is available to be used elsewhere. Imagine that you're not alone, but that there's hundreds of thousands, millions of people doing the same thing as you, with the possibility compensation and exchanges between everyone, depending on their needs. Like the internet, but with energy. And finally imagine that the interconnection of production is done through the city's operating system, the City-OS we talked about a few chapters ago, through which all the city's buildings, all its homes, are interconnected in real time balancing sources and consumption. Nowadays we don't talk about chimeras, but about realities. A city's Smart system can be connected to public and private buildings' Smart systems.

In recent years we have had the fortune of being able to see how energy storage technology has evolved. Today, a battery the size of a small fridge connected to a wind and solar powered system can provide enough energy to power a building for four families. You have got to go and see it in China, for example, where the technology associated with batteries has evolved very quickly. But you don't have to go so far: Poblet lights up the access road to the monastery with designer one hundred perfect self-sufficient lampposts. This is comparable to the city of Shenzhen that uses wind turbines attached to lights to ensure energy supply.

Democratic access to energy is one of the keys to progress and freedom, in the First World and in the developing world. Access to energy means being able to heat water, to heat a house, to power a computer or the WiFi that will allow us to sell our handmade products to the world to pay for our children to study. Professor Solomon Darin, of the University of California, Berkeley, has launched a Smart Villages project in the Indian state of Andhra Pradesh. Intelligent villages connected to the internet where local craftswomen, the producers of wonderful saris and earthenware that would fetch a good amount of money in Chelsea or Ensanche, are present online for the first time.

Smart Cities, WiFi, Hyper-connectivity . . . all these words become small compared to the possibility of an elderly woman, with little in the way of a pension, being able to pay for her heating and the stove to cook her food. A new state that wants to be at the forefront of modernity cannot ignore the energy debate. It's clear that self-sufficiency won't be achieved within two terms in office. What is clear, however, is that if the commitment is all-inclusive, if it encompasses all age groups and all ways of thinking, we will be able to tap into one of the most important known economic sectors.

Very important energy companies are beginning to consider the extent to which it is wise to continue depending on a productive input as volatile as oil. The same big industries could be starting a shift, understanding that in a hyper-connected world (by pipelines as well) there could be a new way to keep warm, to move around or to work. Consider, if not, the possibility of filling up miles of natural gas pipelines from North Africa with hydrogen. The pipeline network and its management could become a new source of business. It would be a completely clean business that would take advantage of existing infrastructure, it would contribute to general well-being and would guarantee a well-distributed supply. The first steps towards energy self-sufficiency take us even close to freedom.

It went by unnoticed. There were no leading articles or analysis. Politicians didn't mention the subject. It wasn't talked about in Parliament's corridors, nor in town halls. The columnists didn't pick up on it either. Antonio Brufau, Repsol's chairman, appeared before the media to talk about the price of oil and gas crisis in the Arab world. After his speech finished they asked him about the price he thought a barrel of oil could reach. His answer was more or less the following, "I don't know how much a barrel could end up costing . . . " He sighed and whispered, "in fact, I'm not sure how the world as we've known it up till now will end up." Brufau had let himself go, something that in his case is not without merit. The aura of mystery surrounding these powerful men, the excellent managers of the great corporations, had disintegrated. For a few tenths of a second Brufau had wanted to share his perplexity at what tomorrow might bring.

FUKUSHIMA AND THE PRODUCTION OF CLEAN AND DISTRIBUTED ENERGY

After what happened in Fukushima, commitment to nuclear energy was blemished. I'm referring equally to the human catastrophe in Fukushima, with its exclusion zone that stretched for many square miles prohibiting all human activity, as to the questions that we dare to pose without being treated like idiots by the inveterate defenders of this theoretically clean and innocuous energy. We're not happy about it, as some are, naturally: but neither do we understand those what want it all, to turn on a light as if it were nothing, but of course they don't want nuclear plants that create waste, or biomass plants that produce smoke, nor wind turbines that ruin the countryside, nor plants with combined stages, they don't want any of it. Neither one nor the other solve the problem, rather they shroud it in mystery. Both positions are backed up with intellectual arrogant logic. We see it evolve in the following way: first they conceitedly talk about theoretically indestructible theories, apologetic for the earth's inextinguishable energy resources. Whenever events show flaws in their logic they get off their high-horses, but not to retract their ideas, but to seek refuge, curled up in their caves of inability to imagine a different world.

The debate on the future of our society must be placed within this context. Today we know that at the current rate of energy consumption in a state of linear equality, that is to say if everyone on Earth had access to the same quantity of consumable entry resources, we would need two planets like ours. We need to roll our sleeves up once and for all and start to think how to solve this collectively, to be able to prosper as we have done to date, and without that prosperity necessarily involving the gradual destruction of our societies. The answer could be energy self-sufficiency, working to build the tools that get us there, city to city, town to town.

Let's imagine any of our cities; for example, the coastal ones. Can you see them: dense, sloping, surrounded by green hills, with a sometimes difficult relationship with the sea that has got progressively better through time, but there are still pending challenges however like the treatment of the beaches, the relationship with the port or the logistic interrelationship of the available infrastructure. Our cities have moved from the industrial sector to the service sector, and theoretically they've closed the door to the way back, towards industry. This has caused the proliferation of tourists in some areas. In spite of everything, one thing has not changed: if we turn off the supply of electricity, gas, or water, not to mention petrol, our cities become useless husks. Thinking about energy self-sufficiency would allow us to design our economic strategy in a different way, it would allow business opportunities to surface where there had previously been none.

Energy self-sufficiency in our cities can pave the way for growth that has been unknown to date. For example, biomass or geothermic technologies; not to mention the energy amassed in the sea, as well as the process of desalination to which we're committing so much, are becoming increasingly more acceptable in terms of the cost of energy. Interventions on the Colombian coast demonstrate this fact, they are finally overcoming the mass diversion of water. Or what we still have to learn about solar energy, or the use of the large green spaces on our coastal mountain ranges, of which we've talked a lot, spaces that are idyllic landscapes in the minds of summer holidaymakers. However they're spaces that have neglected their productive and complementary potential to offset man's activity. The arrogance of a few learned people goes against our collective future. It clings to the past and serves those who are scared by or do not care about the future. They're a minority and more will join them if we dare to consider the present as if it were already tomorrow.

In fact, the future is in our classrooms in our cities' schools. Not long ago, two students in year 13 at Barcelona's Institució Cultural del CIC presented their investigation: "A self-sufficient El Turó de la Peira." El Turó de la Peira is a small neighbourhood in the northern part of Barcelona, built during the years of Francoist policies of economic development, when Barcelona needed to house tens of thousands of immigrants, the majority of whom were from Spain. The neighbourhood is made up of low-quality housing, largely affected by aluminosis and situated in an area that was previously marginal but nowadays has enormous potential: sunny, in contact with nature, with reasonable public transport and a network of facilities comparable to the rest of the city.

The objective of their investigation, Mar and Elisabet told us, was to design a neighbourhood that was energy self-sufficient using renewable sources, without altering its basic structure. They also wanted to demonstrate the installation of renewable energy and energy self-sufficiency systems to improve the quality of life of residents. El Turó, as I was saying, is one of Barcelona's poorest neighbourhoods, with a very strong personality due to the overwhelming growth of our country's capital during the third quarter of the 20th century. It's a neighbourhood that suffered like very few others the effects of planning irresponsibility and the needs created by the economic crash without democracy, paid for by way of demolitions and the suffering of many innocent people. However, El Turó is a happy neighbourhood, it's sunny, it has woods and is full of life. A neighbourhood that knows about shops and workshops, bars and living rooms. About 16,500 people live here and household income is around 60% of Barcelona's average. The domestic energy consumption of the neighbourhood is around 24 GWh/year, whilst the tertiary sector, that is to say services in general, consume about 43 GWh/year. Mar and Elisabet faced a major challenge: to make this neighbourhood in Nou Barris produce all the energy it consumes. The students explored the available technology: from eolian, using wind turbines for cities, to solar and kinetic energy. They designed the installation as a wind farm that would produce around 12.000 KWh/year; they equipped the neighbourhood with an impressive kinetic energy installation which consisted of harnessing the energy generated by pedestrians whilst they walk, achieving almost 12 GWh/year, far more energy than is needed to power the neighbourhoods lampposts. Finally they provided solar panels for buildings and public spaces, sufficient to generate 61GWh/year. The only obstacle they encountered was the impossibility of storing the energy produced by the neighbourhood. Current regulations don't allow it. Mar and Elisabeth smilingly asked themselves what would happen the day when, instead of a couple of sixteen year-old students, politicians scientists, experts, public servants and residents in general decide to go down this route. Think, they said to me, what will happen the day when buildings, blocks, neighbourhoods are able to produce and store the energy they generate in a clean and renewable way, to consume it when they need it or to sell it to the areas in the city that haven't made the change yet.

This brings us back to Japan. It is reported that a Japanese brand of vehicle, Nissan, was on the verge of completing the first mass launch of a completely electric vehicle. The vehicle's specifications stated that the battery could be charged as well as plugged in to "transfer" unused energy. That is to say: a car parked in a garage could be connected the home network and supply it with energy for more than two days. All of a sudden the world's first mass model of distributed energy storage was right before our eyes. A model comparable to the internet and data, but in the field of electricity. When the Japanese energy lobby found out what this implicated, it pressured the country's government to make the brand modify the vehicle's technical specifications and to cancel the possibility of transferring energy. The government gave in and the vehicle was put on the market with specification capped. A few months later a giant wave was drowning the western coast of Japan. The powerful tsunami inundated everything in its path. We are all aware of the consequences, the most notorious being taking the Fukushima nuclear plant to the limit of explosion. It has been said that the government called the directors of the car manufacturer a few hours after the accident and told them to immediately recover the specification that allowed the cars' batteries to transfer energy. The motive was clear: a battery could save the life of a family if the supply of energy from usual sources were to fail for a few hours or days.

This Japanese account ties in with the strategic commitments made by the big automotive brands: on the one hand Volkswagen, who once opted for latest-generation thermic engines that had very low consumption, high energy efficiency and low pollution. On the other Nissan and Toyota, prepared to strategically change the sector, committed first to a totally electric vehicle, and second to hybridisation. The facts have ended up proving the Japanese companies right: Volkswagen, caught up in its own mistakes, has decided to move to the world of hybridisation and electric engines: vehicles that don't pollute, are silent, and have identical transportation capabilities to those equipped with thermic engines. That changes everything: logistics, overground public transport . . . We need to commit to fleets of electric buses. Articulated buses with one, two or three sections, with competitive ranges and high value public service. Overground public transport networks that are flexible and adaptable to real demand, thanks to the creation of algorithms that streamline routes and services, just as cities like Sant Cugat, in Barcelona's Metropolitan Area are proposing. Smart energy leads us directly to transport and Smart logistics. That's the logic behind a hyper-connected city, behind the mantra, and behind the City-OS.

Elisabet and Mar's work, and the Japanese account, offer us clarity about the road we need to go down. Energy self-sufficiency, associated with distributed energy storage and production, clears a new path for everyone. When the youngest members of our society prophesy through the work they do in education centres, and when societies as consolidated and serious as the Japanese abjure the oligopolies that are not able to respond to the crisis, when a small battery plugged into a house's network can share power, it shows a new world opening up before us. There is the way of darkness, which is to deny the prophecy. And there is the way of the light, which is to come to terms with it as a national challenge and set out to be the first to turn it into a reality.

As I have pointed out above, the interconnections between electricity and the city require me to dedicate a few moments to the genuine clean and flexible collective transport revolution: the electric bus. Now that some cities are being filled with train-lines, as if we were on the verge of the 20th century, staring at it from the 19th, instead of realising that we're at the height of the 21st, it's imperative to cry out in favour or more effective, flexible and clean overground transport than currently exists: the electric bus. Everything else is less flexible, more expensive, less adaptable, less reversible, less scalable. In your city you must pull out all the stops to promote investment in overground public transport, with separate lanes or not, depending on the road's purpose and the optimal speeds for each of them. In any case, these days there is more than enough installed power and battery technology to unveil an electric bus transport system that could adapt to the city without compromise. In other words, technical evolution and technological revolution allow us to already be able to think about a method of mass overground public transport that we can adapt to the number of people that use it, but also to the city's changing morphology. No works, no diversions, no reconfiguration of superblocks, no reconceptualisation of public space that are going to be an obstacle to keep adapting bus routes to the changes that the city will experience. Buses that stop at intelligent stops which inform passengers one by one, which identify them as they get on the bus, which let people know about events in a personalised way. I'm not talking about the future. I'm talking about a present we can demand.

If we add to that the fact that we will soon have efficient energy generators based on alternatives to current energy sources at our disposal, or generators that simply make eolian and solar energy spread across our cities in a non-experimental but structural way, we can be sure that the electric bus, as well as the electric lorry, and the electric tourist vehicle will be best commitment to the future that we can make for our cities. The Smart mobility option.

11

Smart Up City: Productive, Hyper-Connected, Re-Naturised

GOOD BYE WORK . . . WELCOME NEW JOBS!

Just before arriving at Philadelphia Central Station in the United States, on the roof of one of the many abandoned factories that can be seen from the window of the train that is taking me from New York to Washington, I see a large, chipped sign that has half fallen down, advertising for the company Good Dye Works. The sign has been modified and now reads Good Bye Work (picture overleaf). It's an installation work by the artist Reece Terris, in which the play on words and the advertisement on top of an abandoned factory take on the role of a 21st century altarpiece.

The work has become the entrance sign to the city of Philadelphia. It must be said that the sign could be stationed outside many of our Western cities.

The television series *The Wire* takes place very close to Philly, in Baltimore. In it, those news Shakespeares that are the David Simons of this world, explain in a brutal and poetic way what we have become. *The Wire* is the television series that anyone who wants to serve a city ready for whatever may happen needs to learn by heart. In one of the many already classic scenes from the series, Frank Sobodka scans the docks in Baltimore port whilst he reflects on what happened to the old port, on where all those jobs went, but above all on the lost dignity of the concept of work, as he and his elders had understood it not so long ago. Frank and his whole family will see how the only possibility left for them is getting involved in the trafficking underworld: drugs, women, luxury goods . . . *The Wire* portrays, as few modern works do, the way in which the demise of the industrial world opens the door to the moral decay of people and entire neighbourhoods. This is the role

Reece Terris, "*Good Bye Work,*" 2012.

of good works of art: to place us in front of a mirror. That's why *The Wire*, and some of its scenes, became our constant reference as we pondered how to recover industrial activity at the heart of our city.

City and job losses go together in a world affected by a crisis diagnosed but untreated, and it does not seem that there is neither the imagination nor the will nor the strength of mind to change the situation. What's clear, however, is that the sign in Philadelphia plays the same role as the old prophets: it removes the veil that shrouds reality from us, making it dramatically visible on the roof of what had been a large factory. Around the factory hundreds of abandoned or deteriorating houses reinforce the scene's drama. It's the image of a Western world that is trying to squeeze even more out of a model that is finished, and it can't seem to find a replacement. It's the drama of youth doomed to failure due to lack of work, of people that end up in precarious or undignified situations when they are laid off after reaching a certain age. All this goes on in cities. The city is the place where we will end up winning or losing our battle with crisis, which is not a circumstantial phenomenon.

In Washington I share a table with the president of the World Bank, Dr Jim Kim, a doctor specialised in the improvement of public health systems. His arrival at the World Bank revolutionised the institution: he proposed turning the Bank into the institution that would set the global benchmark in the mission to end poverty. He is not naïve and knows that this kind of statement can become a very bad advertising slogan in a short time. For this reason he proposed focussing the Bank's actions on initiatives that generate wealth, business and industry, in a clean, efficient and equitable way. This is where the *City Protocol*

http://cityprotocol.org/

becomes relevant. We launched it from Barcelona; it was the reason why the Bank started conversing with us, they even installed a team in our city to work with us in the field of intelligent urban development. What interested them most about Barcelona's approach was the generation of new economy in cities, through the commitment to Smart Cities. They thought that cities were obsolescent living machines that needed to be updated. The spheres of action were diverse: in the field of energy the time had come to produce electricity in a decentralised way; we had to make our buildings start to generate the majority of the clean energy they need to be able to function, without having to pollute the atmosphere. We had to, and we still do, equip buildings with batteries that accumulate the energy we don't use. And for all of this we need to create a new industry that we don't have, that means job creation, generation of wealth and of hope.

We could say the same about the field of water management, of waste treatment cycles (waste to energy) or about the field of the introduction of electric vehicles, where, incidentally, we play a key industrial role by being practically the only city of reference that has a factory that is totally committed to the mass production of electric vehicles. There is also the field of open information management: under the Mobile World Congress and the Smart City World Congress, Barcelona launched Apps4bcn, a space on the internet where the creators of mobile applications related to our city can exhibit and promote their App inventions. We then thought that if the thousands of creators, that is to say, small-scale entrepreneurs that we had and still have in Barcelona, could rely on the city's big data to get new applications off the ground, we would see their productivity and generativity multiplied.

CITIES' NEW ECONOMIES AND THE CREATION OF JOBS

Cities' new economies should force the Philadelphia sign to be changed for one that reads *Hello Jobs!*, a sign that welcomes a new generation of jobs. We began to count on important allies in this war against deference that leads our people to lose hope. We know that if we carry on down our current path of renunciation we will always be at the mercy of others. Because the principal mission of a government is to produce conditions for the society it serves that generate enough wealth so as to be equally distributed among everyone. This wealth will be distributed in the form of education, social services, security, cleanliness, affordable housing, health, transport and all sorts of infrastructure . . . The generation of these conditions is never a product of tactics, nor demagoguery. The conditions for the generation of wealth are related to the productive and competitive faculties of a society, with a good reading of the competitive advantages that place a nation, a city, ahead of others. They are, therefore, the product of a strategy.

Take for example a city like Barcelona, where many years ago someone decided that industry had to originate in the heart of the neighbourhoods. In fact since the end of the eighties until just a few years ago, Barcelona worked to remove all available industrial land outside of the Zona Franca, to turn it into residential land. This decision has had dire repercussions for many businesses and employees, and equally for the quality of neighbourhood life. Business owners, faced with possible expropriation, or turning their factories into flats, decided to stop investing in Barcelona's factories and stopped hiring labour.

Things are changing however. We've succeeded to position the concept of a productive city in all the ideological cores that matter. We decided to urbanistically vacate all of Torrent de l'Estadella's industrial zone, a decision that consolidated thousands of jobs and will give the current industry in Sant Andreu a future. It's one of the decisions I'm most proud of. I'm an industrialist, I'm productivity focussed, I'm a child of the assembly line, of Marxian surplus value, and of the 6 a.m. shift.

BARCELONETA: FROM AN INDUSTRIAL CRISIS TO A PRODUCTIVE SERVICE INDUSTRY

The paradigmatic example that places us so close to Baltimore is the district of Barceloneta. Here the *Maquinista Terrestre y Marítima* factory was opened in 1856. Very few people remember the weight Maquinista had in Barceloneta, and the city as a whole. The turbines that moved a large part of the Catalan textile industry were produced in the Maquinista factory, so was the wrought iron that was used to construct the San Antoni and Llibertat markets. We could also add docks, bridges, material for war during the Civil War, locomotives, thermal power stations and all kinds of infrastructure that was fundamental for the development of the country. In 1965 the factory in Barceloneta was converted into a warehouse, and a few years later it was sold and dismantled. The thousands of workers, residents of Barceloneta, lost their jobs, but also their future. In Barceloneta sudden deindustrialisation annihilated the true personality of a neighbourhood that, as much as or more than for the sea, lived for the noisy assembly lines, milling machines, shifts, moulds and gears; a neighbourhood that wore blue overalls with hands stained by grease. This all disappeared to create a social desert that caused a lot of pain in the form of marginalisation, delinquency, drugs . . .

In the eighties and nineties the city committed to beaches. And with them tourism, restoration, commerce and apartments rented by the day. We immediately noticed the need to reindustrialise Barceloneta, giving support to what remained of it in the old dockyards and barrages. We relied on the spectacular business venture Marina 92, on the reconversion of Marina del Port Vell, on the installation of Desigual in the neighbourhood and the possibility of consolidating the nautical sector of the city. Hand in hand with Port 2000 and key people like Sixte Cambra, José Alberto Carbonell, Toni Tió and so many others we launched the Clúster Nàutic and we turned it into the most competitive in the Mediterranean. For this reason the industrial commitment to Barceloneta, through the Clúster Nàutic, is so important. For the first time in two generations we returned to openly committing to productive industry, the most capable of creating a future and the capital gains that are necessary to transform everything that we produce between us into distributable wealth. This cluster of businesses are carpenters, painters, mechanics, sheetmetal workers, designers, upholsterers, engineers, quantity surveyors, architects, clerks, sailors, cooks, economists, teachers, researchers, computer technicians . . . The cluster is synonymous with jobs, industry and future.

I always say that one of the best things to happen in my life has been the opportunity to work side-by-side with Toni Miquel, nom de guerre Leslie, or *l'Anxoveta* [the little anchovy], the singer of the legendary band Los Sírex, the king of European pop-rock bands. Leslie was a minister of the Barcelona City Council and is a real specialist on the city, one of the people who best knows its ins and outs and secrets. Leslie is an artist, and therefore a sensitive man, close to people and endowed with the popular intelligence that belongs to those who have made themselves. Leslie was instrumental in the rolling out of metropolitan city-building strategies between Les Corts and l'Hospitalet, but also in thinking about how we would move forwards in a Barceloneta that was plagued with tourism due to the logical success of the beaches that had been created behind it, years before.

One of the memorable days of my time in the Council was when we discussed Barceloneta's future with some residents' organisations, and not exactly the closest one to us. Leslie, Albert Civit, my irreplaceable right hand in the Council (and beyond) and I turned up at the meeting, representing the institutional side, and three people on the side of the neighbours: Josep, who had worked at the Maquinista factory; Rosa, a representative of one of the organisations, the same age as Leslie; and Gala, a neighbourhood representative who was younger and more battle-hardened. Gala seemed to call the shots in the group. She addressed me with with a tone that lay somewhere between contemptuous and aggressive, as if someone were talking to their own prejudices. No-one knew anything about me, my family, or my social background, but they referred to me as *powerful, bourgeois, executive* . . . She defended the Barceloneta *of before*. "Of when?" I asked her without malice. "Of before, when there were fishermen," she stuttered. Rosa and Josep lowered their gaze. Perhaps Gala hadn't lived in Barceloneta for long, perhaps she had misunderstood what Barcelona is, but it was clear that Gala was confusing her romantic idea of Barcelona with the real Barceloneta. In the middle of the conversation Leslie turned to Rosa:

"Rosa, how long have we known each other?"

"For more than sixty years, Toni."

"What an old man I've become, Rosa, but you're the same as always . . . " he said, smiling. "Do you remember what a great time we used to have? Flat out at the factory, shops, parties, bars and work, work, work . . . "

"Yeah, you're right," she said. "But that was all lost in the seventies, remember? When they shut the Maquinista. Everything went to shit then." Josep nodded.

"True, because there was about four fisherman, including my dad, but more than two thousand people worked in the Maquinista, two thousand families that were left with nothing."

"You're right . . . and our kids . . . "

"What happened Rosa?"

"Our kids' generation, the ones born in the sixties, was lost to desperation in a neighbourhood without meaning, they were marginalised."

"After being the neighbourhood of Salvat Papasseit,[21] the neighbourhood of dignity," said Leslie. "But with the Olympic Games progress arrived: they tore our beach bars down and almost killed poor Bernardo Cortés.[22] Progress arrived in the form of beaches, tourists, you remember, right? The Olympic Games were going to save us!"

"But it wasn't true . . . " said Rosa seriously. "The Games were a capitalist fraud."

"Don't be like that, Rosa," said Leslie. "It's got a lot better for everyone since then: the neighbourhood is nicer, we can see the sea, there's more activity . . . the tourists get on our nerves but it's true that thirty years ago we were off the map."

---

[21] A Catalan avant-garde poet, considered one of the most important Catalan writers of the 20th century.

[22] A writer, singer-songwriter and actor famed for singing and playing the guitar among Barceloneta's terraces.

"But we have to do something, Leslie. Otherwise they're going to kick us out," said Rosa.

"Exactly. You're right. That's why we invested in the Clúster Nàutic, in the nautical industry, Barcelona's new Maquinista, whether people like it or not," said Leslie watching as Gala sent tweets on her smartphone.

"So the nautical thing is an industrial commitment?" Asked Rosa.

"Obviously," Josep interrupted. "We need to make sure the nautical industry stays in Barceloneta, Rosa. It's the best legacy for our children."

"That's what we're doing," concluded Leslie. "Recovering the industrial soul of Barceloneta without giving up tourism, but getting the best out of it, That's our strategy."

The conversation between Rosa, Josep and Leslie, whilst Gala tweeted and I listened attentively, is one of the great lessons I have learned in life: the city must be the productive corollary that allows for the creation of future. Closing the Maquinista imposed thirty years of pain on a neighbourhood that was too fragile to deal with the crisis in a constructive way. The arrival of the Olympic beaches accentuated the split, the social rift. Only the arrival of nautical industry, and its subsequent combination with the tourism sector, has given Barceloneta, perhaps one the neighbourhoods with most personality in the world, part of the future we need.

HOW TO RECOVER THE INDUSTRIAL HEART OF THE CITY?

The challenge for your city, in the coming years, will be recovering its industrial heart, and if it didn't have one, you'll have to find it through the digital revolution. Smart hyper-connectivity is the birthplace of this deindustrialization, a creator of jobs and opportunities. There's no turning back on cities' commitments to industry. The reindustrialisation of cities is the Modern World's real challenge. The model of a productive city, that works to make itself self-sufficient and clean in terms of energy, is the city model that generates new forms of wealth, in the heart of its neighbourhoods. Without this generation of wealth, there's no way of counting on an administration capable of dealing with the social challenges we face today.

We will also have to fight alongside all governments, from the Catalan, to the Spanish and European, to make sure the value added tax (VAT) generated by commercial activity in a city, stays in the city. The tourism industry brings benefits, but generates externalities that need to be managed by public resources. If the neighbourhoods that are most affected by this industry cannot use the taxes this activity generates, conflicts arise. Productive industry, combined with a responsible and quality tourism industry, with taxes, including VAT, remaining in the city, to provide first-class education and social services. A whole programme of work to generate and distribute wealth.

We're in Beijing, in the headquarters of one the world's most important infrastructure companies, and we stop in front of a large picture in which Hu Jing Tao (General Secretary) is giving the last speech before the members of the Communist Party of China's leadership. The company's president tells me that it was one of the most important speeches in recent years. Hu handed over the country's leadership to Xi Jinping, and he did so highlighting China's main challenge: to match economic growth with growth in the quality of social services, the improvement of education and the recovery of Chinese culture and identity. In fact, a long-time member of the party tells me that China is a carriage that moves about on an axle with two wheels. The left wheel is the economy. Deng's dream of not accepting China as an ungovernable country and a society incapable of creating collective wealth, is the carriage's left wheel. This wheel has been turning rapidly for years, and one can longer speak about the differences between the Chinese from Hong Kong, Macao, Singapore, Taiwan, Hangzhou or Shenzhen. The Chinese are not a people that has been predestined by Confucianism, nor are they a people that only understands the language of orders and reprimands. They are a people, a nation on the move, that fights literally to escape severe poverty, and they're doing it with great success. Just spend a few minutes in Wuxi's new train station. Go to understand where 21st century Liverpool Street Stations are, to understand how far the Earth's centre of gravity has shifted eastward. To see how China moves with logic. Near the station enormous asparagus-like residential buildings tower above what used to be fields; the streets can barely keep up with the rhythm of construction, basic services arrive as best they can.

Without doubt, the left wheel of economic growth is turning with great velocity. What China has discovered, however, is that there's another wheel stuck to this one by an axle. The carriage cannot stay up on only one wheel. It's a wheel that they hadn't maintained until very recently. It's the wheel of culture, the wheel of nondiscriminatory education, of healthcare for all. It's also the wheel of good urbanism, of good architecture; the wheel of memory, of parents and grandparents. It's a wheel that turns whilst waiting for literary and cultural heritage, environmental and intergenerational sustainability. It's a wheel that turns far slower than the other. The old party-member stops his explanation and looks at me. He takes a deep breath and asks if I've ever been in a carriage. I say yes and he smiles, incredulous. Well you know what happens when the left wheel turns faster than the right, then? The man positions his flat hands in front of his chest and starts turning them to the right. "That's what's happening to us in China, and we don't want it to happen. We want the right wheel to turn as quickly as the left. We have to accelerate this one and slow the other one down a bit. Do you understand?"

After quite a few visits to China, and many hours of observation, work and conversation, I've arrived at the conclusion that what I most admire about the Chinese people is their ability to turn thought and reflection into action. They breathe an atmosphere of productive capacity housed in a model that can't be seen anywhere else on Earth. Reducing the time between thought and execution is the key element for success in the Modern World. It was in the old one. This is what Alexander the Great must have understood of Aristotle, his teacher and mentor. He must also have soaked up Aristotle's great love for cities, which for him were humans' natural communities. Hence his passion for cities, his passion for founding them, in only ten years . . . Alexander died aged thirty three. Mozart wrote *Cosí fan Tutte* in three weeks. We took two years to rezone some land. This can't be.

THE PRODUCTIVE CITY

This is the bad thing about Europe: we're stagnated. We think things through until exhausting them. Sometimes we stop for fear of thinking. Sometimes we use bureaucracy as an excuse to stop everything. We're the kings of prognosis, the kings of endless debate. We postpone decisions until the the idea, or the solution it will bring about, becomes obsolete. The great Chinese lesson is in decision-making, the ability to be resourceful, the product directed towards a shared goal. That's why I'm convinced that, despite having to deal with great political, social and political problems, with losing touch with their roots, with communal pathological conditions, they will be able to correct the speed of both wheels to make them turn as in sync so the carriage travels in a straight line.

Some say that only under authoritarian regimes like the Chinese system is it possible to be decisive. It's not true. India is the perfect counterexample. Things get done in China in spite of their authoritarianism. And it's surprising that we haven't realised that our liberal democracy is the best way to make things happen, if we know how to make the most of it, and we don't turn it into a madhouse, in which a few gurus confuse some people's legitimate anger with the eve of representative democracy. An old Marx used to say that Hegel remarked somewhere that all great world-historic facts and personages appear, so to speak, twice. And he continues, "He forgot to add: the first time as tragedy, the second time as farce." Something similar has happened with the people that confuse the age of social media with the age of post-democracy, and it shows.

We'd like to think that the carriage metaphor is transferable to our country. Despite all the regrets, we're moving at high-speed in social, cultural and identitary terms. We know what we want. However, I think we should be more courageous when it comes to addressing the left wheel's challenges, the economy wheel. The advancing world, China, India, are asking us to make better cities, more intelligent, clean, free societies. What are we waiting for to get on the plane?

Let's go to the productive city, let's return to that city full of workshops, warehouses, places in which creativity, craftsmanship, profession and work ethics made everything possible. The city as a total community for human beings. We need to go back to the Aristotelean polis. My father recalls how not so long ago, in Barcelona's Gothic Quarter, everything was full of workshops with rectifiers, presses to make moulds, shelves full of solutions for construction or the metalworking world. When that world disappeared, the city disappeared. Today, the only way to reenergise the city in the places that have been abandoned to the fate of tourists, is the hyper-connection of the Internet of Things, the IoT, with the advantage that digital technology is clean and silent. Because if we want to the city to return to those streets, it is essential to link its medieval alleys to the 21st century by connecting everything to everything. In Barcelona we have a great example, led by a Chinese woman with an intellect as sharp as a knife. I'm referring to Cecilia Tham and her Makers of Barcelona, or MOB. Cecilia has returned the city to the city, making the process of productive return tangible, a process we can use on our streets and squares. It involves understanding that collaboration, community work, the management of shared spaces, and even the collaborative dissemination of knowledge, are not at odds with the generation of wealth, competitiveness, the establishment of business projects, the financial world and knowledge put at the service of opportunity, but quite the opposite.

In your city you must promote one, two, three, ten MOBs; you must look over the map of creators, the map of digital nomads who seek and search for places to manufacture; you must locate the Cecilia Thams and help them, without stifling them, without expecting them to tell you you're right, because they are your *raison d'être*. Palma de Mallorca, in the hands of its mayor, Toni Noguera, has a plan to convert disused buildings into authentic channellers of the digital revolution. Toni had a vision similar to ours in Barcelona: it was possible to combine creativity the generation of industry, and it was possible to do so through the appreciation of craftsmanship and local production. In the old GESA[23] building, that was on the cover of Antònia Font's legendary album Lamparetes, Toni and his team are working on the creation of the Centre d'Experiència Digital: culture, digital prints, creativity, knowledge, universitat, craftsmanship, community and start-ups. A Smart all-in-one, that's already a model. The same thing is happening in Las Palmas de Gran Canaria. There, with a vision of urban and social regeneration, that is integrated in the island's process of digitalisation, the Council has chosen to align the recovery of the emblematic Las Canteras beach with the drive of information society, the installation of training centres and the dissemination of the digital economy. This will be done with the involvement of the University, the Corporation for the Economic Development of Gran Canaria, civic organisations and the most important sporting organisations. Social

23 A gas and electricity provider in the Balearic Islands that was absorbed by ENDESA.

Urban Habitat, that's what they preach in Las Palmas. It's no wonder Las Palmas is one of the European capitals for digital nomads: attracting and retaining talent through an integrated approach to the city's regeneration, that's the key.

WHY HYPER-CONNECTIVITY?

Las Palmas is and wants to be more connected. Hyper-connectivity is not a fad, nor is it a few freaks' snobbery, eager for business cases. Hyper-connectivity will allow the development of new public services, hitherto unimagined. It will allow us to connect the social needs of the elderly to municipal services' available care; it will allow us to link social welfare to health. We will be able to monitor our lives by cross-referencing data such as pensions, housing costs, average monthly outgoings, relationships with family or friends, trips to the doctor's, physical exercise and mental activity. Hyper-connectivity will allow us to correlate traffic with pollution, noise, personal health, the health of the plants around us, with available parking, with available space in shared cars, or any mode of public transport. Hyper-connectivity will allow us to link street-lighting to safety, identity recovery, reduced consumption, energy self-sufficiency, distributed production of energy, sustainability, the expansion of the electric vehicle, paths for schoolchildren, and augmented reality in the fundamental routes for the explanation of what our city is. Hyper-connectivity will allow us to define public space according to citizens' preferred use for it, and not based on the intuition of a few people. We will marry it to suitable urban furniture design, adapted to all ages, to the configuration of new ways of opening up the city for works, to meeting and interacting with neighbours, to the establishment of civic centres for 3D printing, to the monitoring of our professional lives, so no-one is left behind by the progress, and if there is any danger of this, we can learn new ways, to receive that helping hand when we hit fifty, when everything seems like it is closing around us. Hyper-connectivity will open doors to sharing literature, painting, music and art of all types, from the most elitist to the most popular; from folklore to cosmopolitanism. It will define personal relationships with services that have been considered to be of mass use and definition to date: let's think about public transport ticketing, with their coded schemes, logical victims of available technology. We will personalise our relationship with municipal services, with our centres of civic activity, and we will do so with everything in the palm of our hand, from our mobile phone. No passports, no identity cards, no Barça season tickets, or passes for the cinema, theatre or MoMA: we'll have everything at our fingertips to be able to know the best way and the best time.

We don't say these things for the sake of it: interconnected data heals. Buildings, data, interconnected people, heal, and they improve people's lives. One of the keys to accelerate the development of precision medicine, for example, the innovation in medicines and processes, and the reduction of general costs is the maximum availability of data by public and private research centres, and by new and existing businesses. In our regulatory environment we have to make data citizens' property. In the last few years autonomous public administration have implemented many developments that allows citizens to consult available data on a public system. In Catalonia the "la meva carpeta" [my folder] concept was developed. We need to go beyond this. Private healthcare providers do not yet integrate their own data with the public sector's. Agreements between public directors and private businesses are not necessarily public and in the age of delegitimisation that we're living, citizens have become suspicious.

Permission to use data is limited to the clinical trial for which it is requested. The advances which are being made only by combining compatible databases are spectacular, since much innovation is based on a very small number of cases. In addition, privacy increases with the number of cases, greater granularity (i.e. country, province, city, district) can be obtained by protecting the identity of the donor. This issue is particularly important with regard to genetics, and a correlation is observed between the level of privacy and the cost of the analysis. The more privacy, the higher the price. In fact there are patented genetic innovations based on the gene of a specific person. Some specialists are beginning to talk about the need to create a citizen data cooperative, in which citizens govern the use of their data and decide collectively or individually on the conditions in which they will allow their data to be used by each of the organisations/projects interested in them. The mass donation of data would make a significant improvement to the healthcare system, but also the transport system, the education system and indeed all of the systems associated with the vital facts of which human life is composed.

The best thing about it is that everything I've written in the last three paragraphs, everything, without exception, is already possible today. And if it hasn't spread on a massive scale it's because there haven't been enough businessmen or politicians capable of taking charge of the city revolution in the way they did when it was the first innovations of gas, electricity or trains. It is necessary to understand that hyper-connectivity brings a new world with it, a large part of which could be monetised, if we understand that we are all willing to pay a little for a substantial improvement in our quality of life.

We must understand the extent to which we have to take the carriage, and make its good wheel spin. We learned things during the big crisis. I don't know if it was of any use to us. In any case I know that we are finally starting to decode the urban genome. We spend our lives asking for good data, and now that hyper-connectivity is available to us, it's time to start using it correctly. I've spoken about the need for the return of productive industry to the heart of our neighbourhoods, to the heart of our cities. Cecilia Tham is an example of this, but there are many more. The problem is that they are still too hidden. The data has to draw out this incredible talent that is already here and now, changing the city for the better. I said that data cures. And it will do so even more if we know how to properly read it, with social intention. In fact, the next step, the most important step we have to take, is turning this digital world into a machine that supplies citizens with public services.

That will be achieved when we start to focus on the sensorisation of our homes for a purely social purpose. This, say, a Social Sensoring Pack, is a kit that is rolled out in homes at a very low price, that will allow us to monitor fundamental elements of the daily lives of people at risk, such as the elderly living alone, or young people with disabilities in the process of becoming independent. The Social Sensoring Pack will measure what we consume, and thanks to an integration platform, the City-OS we have said so much about, the pack will read our domestic consumption and cross-reference it with our neighbours', and so on until we can manage the neighbourhood or city as a whole.

Economic activity in the 21st century no longer needs functional specialisations. The exchange of ideas, connectivity and collaboration are enemies of zonation. For this reason taking a systemic approach to the city requires all of its organs to have advanced, mixed and combined uses. The difference between doctors and those who devote themselves to cities is that our possibilities are far more open-ended. Stopping a city's inertia is only achieved through plans and projects, and this was another of our well-known secrets. Our Smart city was conceived at street level, in the urban project. Each city does what it can, and the most advanced integrate the most advanced technology to make correct diagnoses and intervene in the most ground-breaking way. We must invest in public space, to improve urbanised environments, and those that could be urbanised. We must reconnect the city's extremities through investment in a new way of creating urban centres. We must do so using criteria that supersede classic urbanism. We don't have a problem with streets and squares, we do however have a problem with an integrated vision for the city, a problem of urban habitat.

A city that promotes intelligent urban habitat, interconnected through the IoT, is getting it right. This is a discipline that's at the heart of architecture, telecommunications, engineering, geography, landscaping, environmental sciences, law, politics, economy and sociology. It's the joint work of professionals from different disciplines that allows legislating, approving and finally building a city that respects and contributes to the development of collective property (the streets, infrastructure, green spaces and public buildings). But also the private, essential partners of social and economic development of society. Believing in this relationship of the positive, collaborative construction of society, between the public and the private, and understanding that the latter is the genuine driving force behind wealth, and the public is the driving force behind equality of opportunity, is fundamental.

But where do we generate a large part of our well-being? To what extent, in spite of being so cosmopolitan, so open and integrated in the world, are we aware of our responsibility? Are we aware of the extent to which a city that openly commits to energy self-sufficiency, that is to say, that jeopardises the environment through the creation of wealth, could become a reference model? And all of this to achieve a democratic and sustainable relationship with the environment, which signifies hope for future generations.

RENATURALISING IDENTITY: THE GATES TO COLLSEROLA, THE CIVIL WAR AND THE SHACKS

Renaturalising identity is not a an aesthetic or useless triviality; but an ethical need, generating quality of life. In Barcelona we committed to developing the "Gates to Collserola" project. The Gates to Collserola reflect the city's shift towards nature, which has always failed us. The idea to take the project forward is the product of many years of discussions at the Institute for Advanced Architecture of Catalonia, of the experience of the work carried out across the country, and the willingness to leave rhetoric aside and act.

Barcelona had created the Olympic beaches, but it had forgotten the mountains. Neither Collserola nor Tres Turons had received the attention they deserved, when in fact they represent very important nature reserves, and when their defining values and the challenges they face by coexisting with urban life, are as relevant as the city's relationship with the sea. The summary of many hours of reflection is simple: we had to move away from the rhetoric of the central metropolitan park and build an effective relationship between the city and nature dominated by man.

To this I have to add the intuitions we elaborated during my time in the Government of Catalonia, a period in which I was charged with redirecting the relationships between the Government, the Board of Trustees of Montserrat Mountain and the Monastery, as well as reestablishing the connection between the Noguera and Pallars Jussà regions. In both territories the biggest challenge was to convince the towns that border Montserrat and those that border the Montsec massif, that we should turn our gaze towards the mountain. It was in the relationship of affection, of identity, but also of *bijective* access between the towns and the mountain where we found the necessary arguments to truly advance towards a redefinition of the relationship between man and nature. In both cases we decided to work with the concept of "gates". In the case of Montserrat, by defining arguments for every town that surrounds the country's flagship mountain. In the case of Montsec, by identifying the pre-existing assets and enhancing the elements that define the massif as a unique place in Europe. The Gates to Montsec and the Gates to Montserrat are at the root of the Gates to Collserola project, already developed by the team I managed in Barcelona. In fact, there are few memories as unforgettable as the days I spent with them in those regions. The country giving meaning to the city. That was my intention when I took the team to those regions. A shot of humility and a strategy to incentivise creativity.

The renaturalisation concept was the driving force behind a large part of the action of Urban Habitat. It was an intentionally humanised renaturalisation. In spite of those who advocated the private renaturalisation of assets, in Urban Habitat we defended the idea that the countryside and the nature of Barcelona's environment should honour civilising action by man, as well as facilitate assets and favour nature being drawn into the city in an organised way. It's still possible to see its renaturalising freshness at

http://ajuntament.barcelona.cat/portescollserola/.

Therefore we prefigured sixteen gates, each generating different dynamics depending on the human realities that had to be protected and in order to guarantee the best quality of life and social opportunity. We did so based on the topographies, formal and informal assets, public transport and the identity landmarks we encountered. Every gate was an opportunity.

From the first moment we were clear that it was necessary to create sixteen simultaneous competitions to develop each gate in sync with the sound of the Urban Habitat mantra: "Many slow cities inside a Smart city." This approach needed to be multidisciplinary, holistic, comprehensive and integrating. The teams that took part in the competition needed to be formed of architects, biologists, anthropologists, landscape designers . . . The quality of the entries was very high. As with so many of the interventions developed by Urban Habitat, the intention was clear: to develop a new city model, based on previous models but improved by the circumstances, exemplary, scaleable, and reproducible. The projects satisfied our expectations, demonstrating the vitality of architecture and city science in our country.

One of the gates that we managed to carry through was in Turó de la Rovira, within the framework of reenergising Tres Turons. Tres Turons is, in the appropriate words of María Sisternas, the city's forgotten Montjuïc. A Montjuïc full of people, families affect the 1976 Plan. We had to act and we did. The plan was to work on reintegration, as we started to do, and on the generation of local identity and civic pride. I mean for the whole city. The restoration of Carrer de Marià Lavèrnia was fundamental. So was the work on perpetuating memory.

For this I relied on a key ally: Joan Roca and his team from Barcelona's Museu d'Història. Everyone worked in order to establish a project that would allow access to the space where the old anti-aircraft batteries are, that tried to protect our city from fascist air raids during the Spanish Civil War.

El Carmel, and especially Turó de la Rovira, is a unique space in Barcelona. One of its seven hills. A central place in the new city. A vantage point that allows us to see the city and the entire country, even as far as the Majorcan Tramontana mountain range. But Turó de la Rovira is also a window to our history, to our identity. Here the Smart is what unites us with the past, with our roots, and what unites us with the future, with what we want to be. Because in Turó the anti-fascist batteries gave way to the shacks of Spanish immigration in the sixties. On these impossible slopes up to three thousand people crowded together. They made from the remains of the batteries and from the terrain that was so exposed to the elements, a home. The Smart city needed to recapture this story using augmented reality, virtual reality and prime public space that was accessible and recognisable to everyone. My intention was to give the place a complete story, the MUHBA[24] and management team of the Projects department devoted themselves to it.

The key is that cities become accustomed to illustrating their identity through the Smart tools that technology provides us. It's incredibly gratifying to see how, through virtual reality, the children of Barcelona can understand how people lived in Carmel's shacks. Not long ago, in a country in the Middle East, camel ranchers tried to explain to us how they worked with their beasts fifty years before. We got by, but having augmented reality tools at hand that could have illustrated the story in a much more direct way, would have been far more effective. Not to mention our cities' old quarters, the industrial districts, rurality engulfed by the city. There is a long way left down the technological path.

A space for identity, for memory, for history, for past, present and future is want we wanted to create in Turó de la Rovira. On a site visit, Albert Civit mentioned how the improvement in the lives of Turó de la Rovira's residents (subjected to the drama of the 1976 General

---

[24] Museu d'Història de Barcelona, Barcelona City History Museum.

Metropolitan Plan's evictions) had meant the recognition of the informal. And with that recognition, dignifying what they were and what they are.

What's Smart, therefore, is renaturising identity. What's Smart is using all the tools at our disposal to redefine areas of reference, the places in which we can all be recognised as a community. We did it in Barcelona and we're doing it in many places across the world. At the end of the day, if Catalans know how to do something well, it's defining identity through the will to be a community. Everyone that lives and works in Catalonia, and wants to be it, is Catalan. Public space and its appearance must imbue the fundamental motto of political Catalanism, in which I have been active since a very young age, and that's the way we did it. Our way of defining what belongs to everyone is so integrative, that today, a few years after completing the works that allow us to climb up to Turó like never before, and that allow us to be there because there's no nicer place in the city to enjoy its views, today, as I was saying, every party, every entity, every local and every tourist, perceives Turó as something that belongs to them. It's the best legacy

Today Turó de la Rovira is a point of reference in the city. The perfect place to see it, to think about it, where youth, history, identity, and memory are combined in the most intelligent way.

and demonstration of the fact that the Gates of Collserola were, are, one of the most important projects for our city, and an example to be followed across the world. Every city in the world has its Turó de la Rovira, in which all types of past are combined, but above all presents and futures. Smart tools allow us to explain this like never before. And they allow us to design public space like never before. It takes intellectual ambition to do so. Look for it. I'm sure you will find allies.

All the city's intelligence, all Barcelona's intelligence, which we have been accumulating since we started to consider ourselves a city (already more than two thousand years ago, it's hard to believe) has no importance if it is not put at the service of the common good, of the social dream that we like to tell ourselves about. We live in a time of unbridled demagogy, a time in which we are starting to be able to employ everything our grandparents fought for. We are living, however, in times, in realities, that were impossible to imagine only ten years ago. From the Catalan Independence Process, to the demolition of the Modelo Prison.[25] From allowing someone to be close to their family and friends thanks to a Smart cities project on social care, to a Manufacturing Association developing a hand for a disabled girl. That's my Smart Up City, that's my social dream for Barcelona, for everyone. Are you up for it?

[25] The Cárcel Modelo de Barcelona, opened in 1904 and built in Bentham's Panopticon style, became a symbol for Francoist repression in Catalonia.

12

What is Technology?

Let's start concluding, and there's no better way to do so than to return first to the classics, and then to New York, nemesis and reflection at the same time of so many things that so many cities aspire to. But even in New York the classics hang around. While in New York we found a copy of *The Jugurthine War* by the Roman historian Sallust (86–*c.* 35 BC) in the Strand bookshop at Broadway and 12th, relatively near to the Odeon where Bret Easton Ellis wrote *American Psycho*. There I was for the first time with that brilliant troublemaker in the worlds of architecture and modern urbanism for social vocation that is the Danish architect Bjarke Ingels. He's someone who has understood that there's no genius without works, nor works that will survive without becoming a permanent struggle to express the spirit of a generation while solving the problems of quality social housing, whilst resolving the revitalisation of a neighbourhood or the energy solution of a community. From Bret Easton Ellis I switched to Sallust thanks to Bjarke and María, singing Manel to herself: "*si hagués nascut a Roma*"[26] whilst walking through Manhattan. Therein lies the power of the city! Well, on that rainy day at the Strand bookstore, we opened the Sallust we had pulled off a lost set of shelves among the twenty miles of books in the shop, and we came across this text:

> The prevalence of parties among the people, and of factions in the senate, and of all evil practices attendant on them, had its origin at Rome, a few years before, during a period of tranquillity, and amid the abundance of all that mankind regarded as desirable. For, before the

[26] "If I had been born in Rome." Track 5 from the 2008 Manel album *Els millors professors europeus.*

> destruction of Carthage, the senate and people managed the affairs of the republic with mutual moderation and forbearance; there were no contests among the citizens for honor or ascendency; but the dread of an enemy kept the state in order. When that fear, however, was removed from their minds, licentiousness and pride, evils which prosperity loves to foster, immediately began to prevail; and thus peace, which they had so eagerly desired in adversity, proved, when they had obtained it, more grievous and fatal than adversity itself. The patricians carried their authority, and the people their liberty, to excess; every man took, snatched, and seized what he could. There was a complete division into two factions, and the republic was torn in pieces between them. (Sallust, *The Jugurthine War*, Chapter XLI, 1 to 5)

Sallust, Thucydides, Homer, Cicero . . . They've taught us everything, together with Llull, Machiavelli, Erasmus, Montaigne . . . Reading the classics involves the following: you stumble across a text and you leaf through it again and again, until you suddenly find some of the keys to the present, to our present. It's as if we suddenly opened a door that seemed closed, which forced us to circumvent the financial markets, the estate agencies, the speculative movements of stock markets, Europe's political inability to confront the most pressing problems that are byproducts of our society's social and cultural changes. *The Jugurthine War* is nothing more than the explanation of Brexit, of that Euro war that has sent entire countries dominated by periods of fiscal irresponsibility to the gutter as if it were nothing. Sallust tells us about the unbridled populism of alter-globalism and the arrogant and dangerous reactionarism that is widespread around the world. In 1989 someone said that the story was over, inviting us to forget. Forgetting everything has led us to all this. No longer do we remember the start of liberal admonishment as a counterpoint to super statism, equally as suffocating. We don't realise the extent to which we have debilitated our poor republican representative democracy for the sake of Pyrrhic victories. And today the city seems to have to face the social, political, economic, geostrategic and even household challenges of the entire world alone.

As a result of all this, it's logical that we hear voices that ask for city parliaments, books that depict a world ruled by mayors, geo-representations of public happiness in which we all compete and collaborate with each other. They are representations of an illusion, which isn't a bad start, but they don't answer the real and pressing challenges that mayors and cities' political and technical leaders from all over the world have to face up to. For this reason I have written this brief invitation to really change things, combining tiredness, feeling, political vision and capacity for action. As the Galician nationalist politician from the last century Emilio Castelar said, first dreams and then ideas, cause historical change. And this is always the result of the combined action of many actors, but always of the action.

During the summer of 2016, the Guggenheim Museum in New York staged an exhibition of a seemingly mad artist: Lazlo Moholy-Nagy. His places of birth and death, both temporal and geographical, explain part of his exciting journey and his practical idealism. He was born in Borsod, a city in the Austro-Hungarian Empire, in 1895. He died in Chicago, in the United States of America, in 1946. Lazlo believed in the transformative power of art, as a vehicle for social transformation, together with technology, for the improvement of humanity. He was a tireless innovator, an explorer of the limits of feeling and life. Throughout his entire life he experimented with all possible media, moving nimbly between the fine arts and applied arts, striving to illuminate the interrelations between life, art and technology. He was an artist, an educator and a writer. It was almost impossible to classify him using the methods of the time. He developed theories and wrote all kinds of essays that influence artists and designers, their echo can still be heard today. Interrelation, interconnection . . . the 20th century or the 21st? Walter Gropius invited him to join the body of educators at the legendary Bauhaus, suppressed years later by Hitler's lethal regime. In 1937, fleeing from the barbarism, he started his American adventure in Chicago, where he opened the current Institute of Design.

Among his explorations in this-art-life-technology unison, what stood out were his camera-less photographs, his use of non-conventional industrial materials for painting and sculpture, his experiments with light, emptiness, space . . . with the objective of repositioning the role of the artist in the modern world and, as such, continuing to make art a catalyst for active reflexion on society. Among the most striking features of the Guggenheim exhibition was the recreation of *Room of the Present* (*Raum der Gegenwart*) envisioned in 1930, but never exhibited during the artist's lifetime. This work sums up a dream that was then impossible: to turn images of reality into reality itself, to explain it; a sensational anticipation of virtual reality, of the digitalisation of analogue, to enter into its substance and understand it better. This is the Digital Experience Centre that all cities should develop. Places in which digital culture is combined with dissemination and entertainment, so that we might all understand it better. Moholy-Nagy was a complete artist, and foresaw things in his time, as great geniuses do. Because the sign of our times is technology at the service of interconnection, at the service of things that talk with other things, of things that talk to people and vice versa. The path from Nagy hasn't been simple, but we're starting to see what in his artistic mind seemed like an emotional ideal, an intuition.

Let's pause here for a moment, and continue in New York: it was around 1870, precisely when Barcelona started to demolish its fortified walls, when a certain Richard Morris Hunt finished designing what had to be the first modern high-rise luxury housing in the city. He did it on Broadway, between 26th and 27th. The building stood out for its height and for the size of the apartments, but above all for certain elements that made it something genuinely new: a steam-powered lift, piping for cooking gas and lighting, hot water, forced ventilation in the rooms, also powered by steam. Stevens House was immediately considered an architectural landmark, it looked to Parisian-style luxury apartments, and was nothing less than a transposition from palatial life to a more social life, a life more integrated in the city, but still disconnected from anything that resembled modernity. For the newspapers of the time, Stevens House paved the way for the foundations of the new city, in which technology had to change ways of living. In an article in *Appleton's Journal* of the time, the editor said that, "We already know something about the European way of living, that is, in apartments, in communal buildings. In our cities some very elegant buildings have been constructed in order to emulate this style of living. But we, Americans, are going to improve vastly in that way of designing and building housing, all of them thought-out before the technological revolution of steam used for domestic purposes. Simply by establishing communications between the street and the highest floors, as in Stevens House, we will be able to give all apartments, regardless of their height, the same living conditions."

What's true is that the project was a disaster in terms of promoting the apartments. Save for a few scholars, journalists and visionaries, very few were convinced by the technological argument. For a few years people carried on constructing buildings with mezzanines, main floors and the rest of it, as if that thing called technology wasn't going to change the way we lived. Of course . . . it wouldn't take more than fifteen years for New York to burgeon as the city of skyscrapers. And in spite of this, we still have our main floors and mezzanines, because we already know that here we always have the ideas cities use to invent themselves. Something similar is true with Smart Cities. We move amongst people who are convinced that the mass entry of technology into the city will save it from the obsolescence we have subjected it to, and those who simply ignore technology, due to incredulity or wilful ignorance.

Smart Cities are already here and it can to be very difficult to stop them. *Àvia* Maria was ninety-two years old, and lived in a first floor flat on the Carrer de Balmes in Barcelona. Just like the majority of elderly people that live alone in Barcelona, *Àvia* wore a tele-assistance button around her neck. The button is nothing more than a call for help in difficult situations. When there's an emergency, the person presses the button and a call is made. If there's no response from the caller or the emergency is confirmed, medial teams or social services will go to the home. *Àvia* had been widowed a few years before, and she lived alone. She wanted to be autonomous, without rejecting assistance; this gave her quality of life and peace in old age. In her case, it was a conscious decision, but in Western cities there are many elderly people that suffer alone. In fact the representation of solitude in Barcelona is an elderly woman who is alone and on a pension, with minimum income.

WHAT IS TECHNOLOGY?

The following conversation took place on a Wednesday, years ago, when Maria used to go and have lunch at her grandmother's house.

"Grandma! How are you? You look good!"

"Really well, my love, better than ever."

"How come?"

"Look," said her grandmother excitedly, putting her hand through the buttons of her shirt. "They've given me the button."

"The button?"

"Yes, Maria, the button. If I fall over, or get dizzy, I press the button and they'll come and get me."

Maria's face lit up almost as much as her grandmother's. "How great, Grandma . . . There's nothing like technology in making our lives easier."

Her grandmother raised her eyebrows, hid the button and shrugged.

"Technology, Maria, what is technology?"

Grandma Maria's button is a friendly voice that pampers Barcelona's grandmothers. Of course it isn't technology! Smart Cities use technology as a tool, not as an end point. That's why Maria's grandmother's innocence is so thrilling. Even John Chambers, who was the president of, "When we read diatribes in opposition to Smart Cities, all of them a product of ignorance and prejudice," laughed with us about what Maria's grandmother would have to say about it. The button's success lies in the collaboration between the public and the private, because it promotes indisputable civic good. Tele-assistance is almost omnipresence.

Gianni Rodari, the Italian educator and writer, travelled often, but he didn't want to give up being able to talk to his daughter in the age of pay-phones. Rodari immortalised the brief stories that he used to tell his daughter in "*Telephone Tales*," waiting for his coins to run out. Today, many of us travel frequently and we talk, cook, eat dinner with or put our children to sleep through Skype or other platforms whose only limit is an internet connection. Hence the evolution of the grandmas' button that stemmed from the *Vincles* project that won the award for the best Smart Cities project in the world. The *Vincles* project's objective is connecting elderly people that live alone with a universe of people that can keep them company, chat and read to them or recommend and prescribe them something with a single tap of a screen. Smart Cities aren't technology, they're social progress.

The combination of the Internet of Things with the digital sensorisation of urban elements, the advanced management of utilities, traffic, public space, and integration with social services' control and management systems, give the opportunity of answering questions that had not been imagined up to now, from an objective point of view.

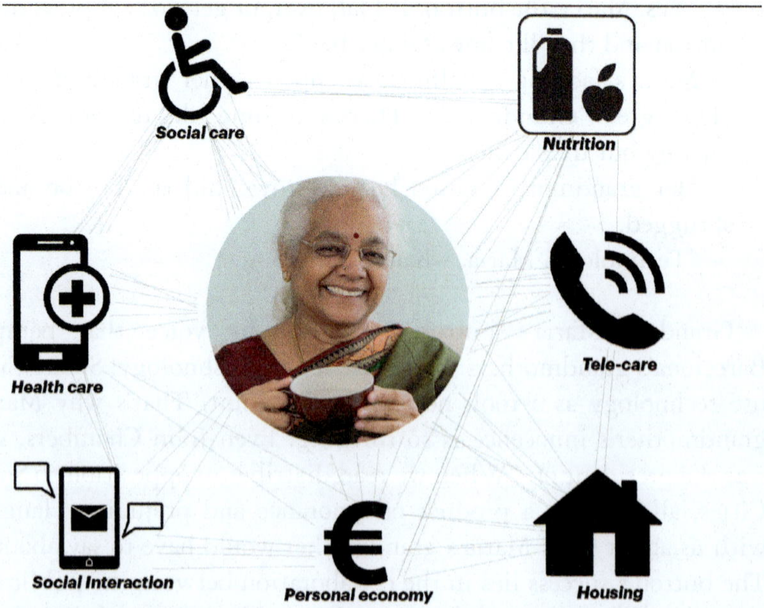

Finally the dream is becoming a reality: to integrate in a single pulse, in a single process, *Àvia* or grandmother in the world's data, by the way, and the data of all the grandmothers there are in Barcelona, Mumbai or Cape Town. Solving a concrete problem and planning the budget for the structural solution to that problem. The simultaneity of individual data and data added in real time is the true revolution of the new governance that is bringing about the Smart City revolution. To achieve this we need to hyper-connect the city, that is to say, everything that can be digitalised, everything, must be digitalised. There should be no geographical or social boundaries. There are no first or second class elements. We are faced with a fourth dimension that has not been taken into account when defining urban services, social services, and the political support of a city, of a polis.

When they tell us that it's not possible to do this in the shantytowns of megacities in developing countries, we raise an eyebrow. It's as if providing water or electricity to these neighbourhoods isn't possible, and people say it. As we were told by the person responsible for the introduction of running water to the poorest neighbourhoods of Cartagena de Indias in Colombia, it's precisely the residents of the poorest neighbourhoods who have the greatest need of receiving good running water at fair prices. They're the ones who fight most to ensure that this water is at everyone's disposal, and controlled by everyone, without exceptions. The same goes for electricity. And the same will apply to the internet and to the urban and social solutions that it brings. Through connectivity we are able to monitor health in which we can find Mrs Carme or Mrs Sita, social situations, as well as MOOCs (Massive Open Online Courses) through which we improve our professional training, or our love of culture. We can also monitor risk prevention, the distributed production of energy, the management of security, and of traffic, the 3D printing of our domestic utensils, our children's toys, the expansion of virtual reality in our phones and the maintenance of our parks and gardens. Technology is the least of it; what's needed is the ambition to transform.

I don't want to finish without listing some of the projects than can help concretise some of the proposals described in these pages, written with a desire to provoke the leap to action. We don't have infinite time to make Western society believe in itself again. Every day of action in favour of a liberal, secular, democratic, fair society that defends the freedom of the individual, the right to commercial activity, rights and duties, is a day won in favour of collective freedom.

I'm writing this a few days after the United Kingdom's decision to depart from the European Union (June 2016). Among the more than sixteen million Brits, (I can't believe it), that voted in favour of Brexit, not all of them are a band of xenophobic neonazis, nor a flock of Eton boys, hiding behind a stupid referendum to get up to their last piece of mischief. In the vote there's a mixture of hopelessness related to the most powerful collective European dream, so well incarnated by the true heroes to the postwar world: Monet, Schuman, de Gásperi, Adenauer, even Churchill, despite so many shadows . . . All of them, men from a Europe based on cities, from here and from there, but rooted in a very powerful idea which is perhaps the best thing us Europeans have given ourselves: liberal social democracy and critical competence; and also the worst of what we've produced, the totalitarianism of all tendencies originated in so much innocent bloodshed. Today we have to fight to make people believe in democracy again as the political space where we can return hope to everyone. Putting an end to the 2008 crisis with a financial sector that has had its debt more or less written off, whilst leaving a hopeless and broken society, is not right. It's the surefire way to brush with one-man alternative political systems, division, robbery and pillaging.

PROPOSED CHECKLIST

That's why the alternative offered by the path towards Social Smart Cities is so powerful: it's feasible, objective, it's an economic sector of its own, its a transformative tool. These are some of the projects that we invite you to consider for your cities. It's a list that, at first sight, seems short, that sets its sights low. It's not true. If you stop to think, if you read it thoroughly, and you superimpose it on your city, we're giving you clues for you to elaborate your own action plan:

1. The creation of the Urban Habitat unit, with maximum political leadership and relying on social and urban transformation, (the project is in charge).
2. A Master Plan for Hyper-Connectivity and Telecommunications.
3. An Operating System for the City and a Platform for Integration of Civic Services.
4. The integration of civic big data and the establishment of a Chief Information Officer and a Chief Data Officer.
5. The introduction of digital innovation in the city's training plans, for all ages.
6. The industrialisation plan for the management of technological innovation (UNE 166002).
7. The digitisation plan for accessible culture.
8. The plan for the deployment of Public, Civic and Licensed Manufacturing Associations (Makers, Fab Labs, Informal Training Centres . . .).
9. The package for the Social Sensorisation of housing.
10. A Social Ties or Bonds programme for vulnerable groups.
11. The package for the Sensoration for Sustainability in homes, office buildings and services of all types.
12. A plan to roll out energy self-sufficiency.
13. Smart Water Management.
14. A Master Plan for Lighting.

15. Urban transformation based on the superblocks strategy, natural corridors and metropolitan hinges.
16. An Intelligent Urban Mobility Plan (Smart Mobility), with a direct leap to electric mobility for overground transport.
17. Acquisition of land for housing development with integrated services, mixed uses and Smart communities.
18. Intelligent urban furniture and the monetisation of urban space.
19. Intelligent urban green spaces.
20. The regeneration, rehabilitation, mobilisation, and digital connection of historic and patrimonial neighbourhoods.
21. The regeneration, rehabilitation, mobilisation, and digital connection of peripheral and disadvantaged neighbourhoods.
22. An Urban and Social Resilience Plan and Government.
23. Integrated waste collection and recycling for energy production (Waste to Energy).
24. An Intelligent Tourism System for visitor-city interaction and the diversification of areas of interest.
25. An action plan to adapt legislation to the new digital reality.
26. Program the "City in Your Pocket" through mobile phones.
27. Implementation of identity on phones (let's do away with cards!) with the maximum possible amount of certification.
28. Plan for the generation of digital social participation platforms.
29. Branding and international positioning to attract investment and talent (the best investment is one that attracts talent to the city).
30. A City Apps Plan.
31. Plan your city in the cloud or Cloud-City-Services: someday the cities of the world will share a database of solutions and civic services that can be implemented without difficulty.
32. A Digital Experience Entertainment Centre (digital education through leisure).
33. A plan for the incentivisation and clustering of digital start-ups.
34. The development of a shared creative strategy to generate a network of cultural creativity.
35. A plan for micro-urbanisations that regenerates abandoned

plots of land in the city, early social victories to set a precedent, and . . .

36. Whatever you can think of as long as it's based on data, has a clear social focus, generates job opportunities, or creates businesses and fits in with your civic mantra.

Although we have reviewed some of the aspects covered by this plan action, each point deserves more than one monograph. What I would like to achieve is that you, mayor, public servant, common citizen, understand as I do that these thirty-five points tells us that we can develop a transformative and feasible action plan, without verbosity and orientated towards fact. We would like to convince the skeptics that technology, well-orientated knowledge, reinforces the stock of our cities' social potential; it makes them better in terms of quality of life, and as a result, more competitive and fair.

As always, however, the proof of everything I have just said is in the pudding. In 2015, 60% of investments in start-ups in Spain were directed to Barcelona, and more than 50% of them were related to the city's new economy, related to the world of digital economy. Barcelona is one of most attractive cities in the world to live and work in . . . according to all the serious world rankings. None of this is the merit of one or several people. It's the merit of a lively, restless, flexible and adaptable society. And many governments. Nothing is achieved without building on the work of previous governments, to whom we have always been and still are grateful.

However, despite all the regrets, today Barcelona is the undisputed Smart Cities world capital. That is to say, Barcelona is the global benchmark for developing the new way of making cities fairer, freer and more accommodating. And this is, and will be essential to continue generating wealth, and sharing it equally. Because Barcelona's got enough left for this revolution, don't doubt it, and for many more.

A Final Comment – Looking towards India

The first time we strolled through Pune, we did so with western eyes, those eyes that see everything as if it were misfortune. They're those tired eyes, somewhat ignorant of what a genuine fight for life implicates. It's not that we have everything sorted at home, far from it, but we can say that we live in one of the most egalitarian, equitable and integrating cities in the world. Pune, as much as Hyderabad, or Mumbai, but also Bogotá, Istanbul or Philadelphia, to give some examples of cities in which we've worked, are completely, or in part, cities that test the human experience to its limits. In Philadelphia for example, in some of the more disadvantaged neighbourhoods, organisations like United Communities,

www.usecp.org,

are responsible for making possible what, at first glance, seems impossible: integrating, offering opportunities, creating a sense of belonging . . . The trouble is that the cultural and socioeconomic environment, let's face it, of racial exclusion, makes everything very complicated. In south-east Philadelphia the city is bankrupt in all senses of the word, and every action has a certain heroic quality. In India, however, everything inspires optimism.

As I was saying, let's look at Pune: through our eyes the city is half-finished, or half falling down, depending on how you look at it. It seems to have ignored everything that we have become accustomed to: organised, thought-out, quality public space; respectable and well-maintained urban furniture; infrastructural services like water, sewage systems or rubbish collection, electricity or public transport, as a basic emulsifier of communal life. None of this is evident in Pune. Either because there are no guidelines for architectural heritage management, or because having been considered at some point, they have now become surplus to requirements.

In Pune there's something that's difficult to understand at first glance: after many years, perhaps hundreds, the majority of the population is starting to realise that they too deserve progress, that they too have a right to keep a part of the extra resources they generate through work, what we call capital, and that it is the guarantee that ensures we all prosper. The *punekars* want these capital gains to exist, and they want them for themselves. The ambition is for these gains to create a sufficiently large middle-class so that the demand for design takes hold in one of the most vibrant and alive cities of the Indian subcontinent. The families that are fighting to generate wealth are starting to demand results from their taxes. They are starting to demand police control. They are starting to expect that those who reject the culture of creating productive economy are no longer given moral lessons, since everyone knows that the alternative is the extreme poverty with which they still coexist.

In Pune, as in a many parts of the world, as it had been in our country until not long ago, hunger is lurking around the corner, and we already know what these collectivising formulas offer, with their undercurrent of trite, quasi-religious morals. Equally, these families, exposed, as they say, to travel, family members who live in more developed countries (although perhaps less free), start to demand collective improvement, that air, gardens, schools, hospitals are improved. That the streets are improved. And in Pune, as in all of India, the engine for change is called Smart City. They are rapidly overcoming technological confusion there. Smart is what unifies traffic management policies with policies for the management of pollution. And everyone knows that in order to achieve this effectively and efficiently, they need sensors, fibre optic, cameras, GPS . . . Come on, it requires exactly the same things as buying a book on Amazon, or doing an online shop because your daughter's cricket match, your son's piano lesson, the party or neighbourhood association meetings have made it impossible to physically go to the shops.

Moohbee is an English teacher at the Mahatma Gandhi state school in the centre of Pune, in one of the poorest neighbourhoods of the city. Moohbee is in charge of children whose parents are mostly illiterate, but know that sending their children to school will guarantee future prosperity. They want their children to be the protagonists of a new Indian dawn. And to do so they need to learn English, as well as Marathi and Hindi. Moohbee has her ideas clear:

to use her mobile phone as a fundamental tool to work with the "Hello English" app.

In Pune the challenge is starting the city, starting it again. In Pune you can feel technology's transformative power applied to daily life. Moohbee embodies this passion for new technology as a catalyst for the search for new solutions. Moohbee's passion in the classroom is Kunal Kumar's in his office in the Council. The challenge is elevating quality of life by creating opportunity, generating wealth, educating and by improving public space and services. Perhaps it's the same challenge as in Europe. It's just that in the case of Europe, the challenge is combined with the explicit or implicit indifference of important sectors which are run by cities' governments, as well as some social and ideological sectors. Meanwhile in Pune, ambition is paving the way for accelerated change via Smart Cities, the sign of predestination. In Europe bad fate seems to have enthralled some mindsets. We can't, we must not give in to it. The battle is profoundly ideological and its final result carries definite consequences for future generations.

Urbanisation, together with other factors such as health and education, is a powerful agent of development. We Europeans have studied it, they've told us about it, or we've intuited it through reading and historical review. The Indians are living it. Pune, 95 miles from Mumbai, three hours away by car or less by plane, on the Deccan Plateau, irrigated by the monsoons that transform it into a green and damp tapestry, in which it's still possible to sow basmati rice just as the painter Joaquín Sorolla depicted in his works on the Valencian rice paddies . . . something we hope to stop seeing as soon as possible. In India, as in another era in Catalonia, the bucolic has always been tied to suffering. Here one understands the extent to which development, economic growth, is a good thing, and how those who are weary and think that they have grown a lot, only allow talk of subsidence.

Pune is around a thousand years old, it's full of history, memories, markets, noise, contradiction, Hindus, Christians and Muslims; of Parsis, Jews, Jainism, Atheism and Buddhism. But above all of boundless vitality. More than three million businessmen and women of trade live in Pune, together with freeloaders, telecommunications engineers, millionaires, soldiers, lawyers, doctors, university lecturers and school teachers. Pune is a city that exudes life, where its density becomes known in every corner. The cars, the rickshaws, the bicycles don't stop, they brush up against each other, they honk, they bump into each other clearing out pedestrians and the jungle trees that almost literally run along the pavements. Pedestrians don't wait at crossings. It would be useless because there are no traffic lights to obey. And the few that exist are merely haphazard guides. The hordes cross the street dodging the stream of vehicles that flows through the city's arteries from east to west. Anyone who has been to India will recall the monotonously exiting chaos that makes their cities beat.

In Pune one experiences the benefits of working in a metropolis in the middle of its creation, the reality of living in a city where basic services that are destroyed, depleted, abandoned, rejected and ridiculed are in the process of permanent reconstruction. And now more than ever, because in Pune it's possible to feel the throb of that magic concept that goes hand in hand with human nature, and therefore with the city: ambition. In this case, ambition for transformation.

In Pune one lives toe-to-toe with the second great Indian transition: the first, in 1948, gave India its independence. The generation of Gandhi, Nehru, Patel and so many others, by putting an end to British colonialism, they put an end to centuries of domination by people. However it wasn't until the nineties that India started the second transition, one that has surely brought it to a crossroads: India is governed by a more open society, in which the social ladder, that has been frozen by the almost invisible but omnipresent caste system, has been overcome by the power of market laws, but above all by urbanism. Large cities are urban equalisers, and India is home to some of the most impressive, intriguing yet also frightening megapolises. The Indians know this, and a few years ago India decided to commit to Smart Cities in a resolute way. The driving force is in the intelligent transformation of public space, the connection of all of its vectors, in sensorisation, the gathering and processing of the available data, the monitoring of the most vulnerable and disadvantaged and the creation of new opportunities. India, with Pune as its best example, has realised that in order to do this it needs to carry out four fundamental steps, that can serve as a blueprint for any sailor lost at sea:

1. Bravely define the vision and objectives, which should always focus on the individual as a starting point and an end point.
2. Implement the best possible model for governance, adapted to the characteristics of the city and its administrative structure,

based on the integrated analysis of the data produced by the city, regardless of whether they come from a public, civic or private source.

3. Work with the best companies, involving them in the development of an ecosystem of local projects and businesses, with the goal of developing public and social services, urban transformation, education and high-quality employment.
4. Develop a programme that involves citizens that involves the criteria for the monitoring and the final evaluation of the implemented solutions.

What interests us about this model is proposing that the speed of urban transformation in India combined with personal and technical intelligence, could reduce the time that First World cities need to reach their level of development. In cities like Pune, everything is up in the air, and everything is possible. Pune could be that affordable city that it seems Mumbai cannot be. The errors made in the first post-independence years in Bombay, especially in the area of housing, freezing rent, undermining the judicial security of the property market, have turned it into a city that is sick at its core. A lack of bravery in turning the development hub Navi Mumbai into a counterweight for the exhausted city that was already the capital of Maharashtra in the seventies, meant that when architects like Charles Correa proposed saving Bombay, they gave Pune a great opportunity. If Navi Mumbai is a commuter city that is struggling to become an integral city, Pune already is. If the monster that is Mumbai collapses from dawn till dusk, shut down by the excessive traffic from Nariman Point to the Airport, Pune remains a human-scale city, when walking needs to become a pleasure again. Today Pune intends to compete with Mumbai with a commitment to quality of life, the incentivisation of productive capacity based on technology, knowledge, professional skills, a clean atmosphere and the creation of quality public space. The future of cities is in the hands of those that design it, and at the moment Pune is in the hands of a creative and able team.

Pune is an eminently young and scandalously optimistic city. It's not the streets overflowing with cars and smoke that make us envy Pune. It's the possibility of change that it can achieve. At the Lighthouse Foundation, which takes in young people that have been left out of the school system and trains them based on their innate talents, we saw how a 23-year-old architect explained the re-urbanisation of the streets to gain pavements and transport corridors. Future businessmen, translators, hairdressers and fashion designers, a few incredibly young mothers of children born in the worst conditions, explained to use how they hoped to change the future that they never had.

It's a profoundly civic, generous and brave society prepared to make operational decisions and share the risks. When Cerdà designed Ensanche, he accepted responsibility for a lot of risks, he put his reputation on the line in spite of almost everything and everyone. He knew that the reality, the attention, or the fact he was an outsider, could put on undesirably twist to his utopia that only made sense on paper. Nevertheless, he didn't renounce the value of projection, and the value of making the jump in urban scale that defines Barcelona today. When Joan Antoni Solans decided to buy land in strategic enclaves of the city in the eighties, he did so thinking about public land reserves in the future. Thanks to this stock we can regenerate, we can reach positive agreements in the public interest with private developers, we can transform and implement impactful public policies.

In the last few years, however, some politicians and all kinds of commentators have taken it upon themselves to systematically see the negative side of development. Private equals bad. Regeneration equals danger of gentrification. Tourism equals profiteering. And so on and so forth to infinity. Not so long ago, Florian Lennert (Director of the Intelligent City Forum, LSE), with whom we have shared tables and debates, answered one of his students enraged that gentrification, *per se*, is not a bad thing. What land-owner from a poor favela in Rio would not want to see his plot increase in value? This endogenous gentrification is what makes society prosper. And if they don't believe this is true, then they should talk to the owners of the block of flats in Barceloneta, just at the end of Passeig Joan de Borbó: from being a building situated at the city's margins, it has become the city's authentic balcony. What certainly is pernicious is urban inaction: what is certainly pernicious is giving up acting in informal neighbourhoods for fear of possible gentrification. This is where politics has to act with all its might. In short, whatever they might say, prosperity is hope, in Barcelona and in Pune.

In Pune the transformation of the city is indeed equivalent to hope. The state invests, and looks for private partners to accelerate the speed of change. It looks for the best. Because it aspires to have a first-class transport system and not a network of dilapidated and polluting buses that loses clients due to their poor condition. The state does it in an intelligent way, because it looks to maximise the service's profits without financially encumbering themselves with unnecessary infrastructure. The Pune's municipal transport company has integrated GPS into all of its buses. Yes, including the old Tatas with rigid leaf spring suspension and windows stained by the smoke and dusty grime for the street. The GPS, linked to a technological platform, allows tracking of all the units, knowing who's driving, what stops it makes and how many people are on the bus. Linked with a payment card, it tracks the origin and destination of all its users, and it keeps track of payment by the minute. Pune will soon improve buses and routes, and it will combine this with active policies of mobility. Politics that generate policies, that's the question.

Time will tell if the bus network improves in Pune. We have to resign ourselves to renouncing a part of the Western policy of working for progress in technology. In European cities there aren't comparable transport problems to those in India, because they don't grow at the same rate as developing cities. However, a rebound in consumption, a return to economic boom and resistance to change can take us back to familiar situations. Will we be prepared to change at the required speed? Smart Cities allow a more intelligent development model, where investments are optimised in favour of what really matters.

In a few years Pune will be able to create a network of buses as efficient as Barcelona's without having to sell land and rezone it. In a context where users will pay more and more for access (to services, housing, consumer goods) and less for property, a development model based on the lending of universal services at affordable prices is a model that will avoid mortgaging a city's public land, that scarce good that once sold is very difficult to get back.

Pune's streets need to be rebuilt. They're wide, because the British designed them using generous standards. But all the space has been colonised by vehicles. Beyond the great transport corridors, that will be dedicated to mass transit systems with dedicated lanes, the network of streets has enormous potential to create all kinds of opportunity. To start, the pavements need to be recovered: to make them free of obstacles, accessible, and interconnect them with understandable pedestrian crossings. They must be wide, so that what are now obstacles, become facilities: the trees and their shade that is appreciated during the endless Indian summer, the bus stops, the street vendors, advertising, tables outside cafés, and signage for guidance. The space needs to be organised to be able to park, because the objective of moving around by car is to stop somewhere to drop someone off, pick up a family member or distribute goods. If the city doesn't allow for organised parking, drivers will obtain it. Pune was the Indian city with the most bicycles until the automobile revolution: now they've been replaced by motorbikes and rickshaws that scrape by next to lorries and buses.

Regenerating Pune's streets for the 21st century requires leaving space for each of these uses whilst taking care of the city's strategic objectives, and aiming to provide the best services. It involves designing an ample network of underground installations, capable of incorporating fibre optic or any kind of new cabling without having to open up the entire street. It involves using lamp posts as a support for lighting and all of the sensors that will collect data on how the street is used. It means that the city will be able to monetise the use of one part of the street for controlled parking, and that bus stops will provide waiting times for buses to the minute.

It's also true that the introduction of technology in the city requires an adaptation period. Designing platforms to pay for parking is useless if no-one sanctions the people that break the rules. Similarly, investing in public transport is useless if you have to go up and down dozens of stairs to take it. A Smart City must be intelligently comfortable for its citizens. Because ultimately, civic participation is also participation in cultural change that entails changing habits to contribute to making public space truly public. To achieve the necessary speed of transformation, developing cities will need to reinforce transformations with municipal support, putting eyes and brains on the street that drive change with the help of technology, not the other way round.

However we look at it, we're coming out of an era that has lasted about a hundred and fifty years. We're living the definitive end of industrial society. The manufacturing methods of industrial and consumer goods are undergoing such a dramatic alteration that it's difficult to imagine how these processes will be in five years time. All of this has important consequences on production processes but also on people's spirits. If not, how can we explain Brexit, or Trump or the Italian Five Star Movement? Globalisation has rendered national states useless. India produces telematic service programmes that produce far more benefits than the production of private cars, let's say. However, the solutions to the country's pressing problems exist on the level of the street, town or city, at most. The world has been changing its axis, and today the Atlantic axis has given way to the Pacific axis, just as five hundred years ago we left the Mediterranean to cross over to the Atlantic axis. All categories of industrial society, as well as our cities, are victims of fervent obsolescence. Cities must recover, or gain for the first time, the status they deserve, physically and politically. For the good of everyone.

Smart Cities, as we understand them, are a commitment to an open society, an inclusive society, one that takes into account the impossibility of building futures with the percentage of marginalised people that the end of industrial society is producing. We opted for Smart to generate the new economy of the city, and with it new sources of wealth and well-being. The Smart project requires a commitment to training and education. Hence our Manufacturing Associations strategy. Hence City Protocol and the global exchange of the best ideas and models. The Smart project also involves, we've talked about it more than enough, listening, listening and listening. That and the new way of participating, generating criteria, analysing diagnoses and making decisions based on all of this. It takes courage, ambition, audacity and intelligence, I accept that. And nothing I have said leads to political success, in the sense of receiving indisputable support at the polls. What it does lead to is peace of mind, knowing that we've worked for the common good.

I was saying that in Pune the word "ambition" is important. So is the word "hope." In Pune the implementation of Smart solutions for the city, and even their use as a driving force for change towards dignity and prosperity, has once again turned into, as in Barcelona, social policy and economic recovery. Build without fear, accept challenges, that's the key. And in India they understand this better than in many other parts of the world.

Pune is making great strides forward. In Barcelona there's a lot to do and most likely a lot to correct, even so we still reached some relatively important milestones, in a very short time:

A. More than 8 billion euros of foreign investment arrived thanks to the correct branding of the city based on the general commitment to Smart Cities and technology.

B. More than 50,000 jobs were created in sectors aligned with Smart Cities and the world of digital technology and general innovation, at the height of the crisis.
C. Ten tonnes of CO_2 and 600,000 litres of water were saved thanks to the implementation of Smart management solutions.
D. We became the European capital of innovation and indisputable benchmark for the movement of world cities.
E. New industrial clusters were generated in the urban planning: Smart Cities, digital content and nautical fields.

They're just statistics, we know. What really matters isn't data or statistics, but the effect of our decisions on the people we serve. *Àvia* Maria, for example. Everything that we do has to consider *Àvia* Maria as the bigger picture, as the start and end of everything that we implement. *Àvia* Maria is Mrs Sita in Pune, or Mrs Fatima in Cairo. What matters is knowing that they're there, and with them their families – that is ultimately what counts, and that's who we serve.

The Smart City, in Barcelona, together with the social dream that has been, is, and will be our city. It makes us better as a human community. I only hope that our humble example, and these even humbler reflections, might contribute to a more hopeful, freer and fairer world.

Epilogue

Adam Austerfield
Director, Global Markets, LSE Enterprise Ltd,
London School of Economics and Political Science

"Dubai, Mumbai, Shanghai, or bye-bye" was what a young LSE student once told me, as how she saw the future upon graduating from the School. These are the Mega-cities, with all the productivity and challenges that come with being the size of many countries, in one urban conglomeration. Antoni Vives is no stranger to any of them, and indeed to many other cities and regions in the world in which he has collaborated with the London School of Economics and Political Science (LSE), often with myself, over the last 20 years. In *SMART CITY Barcelona*, Antoni makes an important, personal and deeply experienced contribution to the field. His approach is of truly civic-centric urban development, allied to a clear vision of what can and should be achieved for promoting human development through intelligent city design and management, instead what is just clever, technologically possible or the latest fad. As he maintains throughout the book, sprinkled with well-chosen and narrated historical analogies, Antoni reminds us that deeper, core values and visions matter when seeking to help transform societies living in urban environments – but that it can be done, and is indeed one of the great challenges of the 21st century and beyond.

We all know the headlines. Approximately 50% of the world's population live in urban environments today, but that will grow to 90% by 2050; cities cover 2% of the earth's surface but consume 78% of the world's energy and contribute 60% of global carbon dioxide emissions and other greenhouse gases; Gini-coefficients are often counter-intuitive in comparative metropolitan areas – London is one of the most unequal cities in the world, second only to Santiago de Chile, for example; and many more assymetrical statistics. But in this book to improve urban living, Toni takes deep dives into many of these global problems with concrete examples that don't involve concrete – building connectedness coupled with smart public spaces that create dynamics that address inequality and its accompanying issues in a sustainable, bottom-up fashion, whilst empowering citizens to make informed decisions about their own neighbourhoods. The clear importance of properly functioning, multi-level governance structures in urban communities is also emphasised, without which no amount of money or construction can create real civic pride in a sustainable fashion. Toni's radical thinking is a noted feature throughout the book, and his plans for 'urban canopies', re-engineering Cerdà's 'superblocks' and the 'City Protocol' are consistently put into the context of thinking about the tangible civic impact of urban development.

One of the great assets of SMART, as much work we have done together at LSE, is that it explores the real challenges and posits solutions through an interdisciplinary prism. A polymath both in academic and professional training, but always a passionate advocate for the power of urban development, Toni draws on sociology, economics, history, architecture, political science, film and even poetry references for a book which gives a fully-rounded vision of how to analyse the challenges of urban transformation and the root causes of many urban problems. The book is littered with clear examples of practical projects from Barcelona and around the world, but also of missed opportunities and failures of how to tackle them, from which we often learn more than seemingly successful short-term hits.

As Deputy Mayor of Barcelona, Toni had the opportunity and drive, in a short period of time, to implement as much of his vision as possible, aided by a remarkable team of talented, passionate advocates for change. These policy decisions and practical implementations were the culmination of deep intellectual quests and robust debates and discussions around the world, not least at LSE, but also an understanding of the power of data in the contemporary age. Here the book does not underestimate the importance of the 'essay questions' – what are we trying to achieve, and why? What kind of data should we collect, big or small? And to what application and policy outcomes can we analyse and use this data for furthering the vision of equitable, progressive society? What role should the private sector play, and how should the public sector manage that role? SMART casts a well-trained eye over these issues, and seeks solutions that fit together and upon which new dynamics can be built for the betterment of society.

But this is no paean to a utopian dream of how things could be; Toni takes a hard-nosed, practical look at what it takes to get things done that make a difference – and stay that way.

As with many aspects of life, long-distance relationships are difficult. Connecting cities around the world through sustainable infrastructure that serves the local community as well as an increasingly mobile international one is a major challenge to ensure some cities do not just become airline landing-fee specialist hubs or tourism ghettos. As detailed in SMART, from his vast knowledge and practical experience of being in charge of Urban Habitat, Toni shows why Barcelona has a myriad of examples of successful inter-linked projects in creating a dynamic, world-renowned city with a reputation and brand for creativity, culture, industry and fun; it has been a long time in the making, but has without doubt succeeded. However he also showcases the problems that such successful internationalisation of a city, not just Barcelona, can bring unless accurately measured and carefully managed data ensure an environment that is focused on well-being for its citizens and provides sustainable outcomes, not just high-impact images.

However in reality short-distance, intimate relationships are much, much harder. Identifying the root causes of social problems in derelict boroughs and poor local communities, where social exclusion, crime, drug abuse and high unemployment proliferate, is not an easy one. But this is where the underlying ethos of SMART belongs – to not forget those who are being left behind and to consider the urban environment in which these social problems manifest. Micro-urbanisations, from Eixample in Barcelona to Pune in India, are given close inspection and support through relative micro-funding, a process through which neighbourhoods are 'dignified', not gentrified.

Finally, Toni's encompassing vision is an invigorating and infectious one. All of us fortunate enough know him well also know that his leadership can transform vision into impact, as is increasingly being witnessed around the world. Mayors, public policy specialists, urban planners, architects, politicians, academics and most of all citizens should get SMART.

Index

About the Author

Antoni Vives is a political consultant and analyst for cities and governments across the world, as well as for the London School of Economics. Between 2011 and 2015 he was the deputy mayor of urbanism, housing, infrastructures, environment, energy, ICT, and innovation for Barcelona City Council. During his mandate the city received numerous awards, including the European Capital of Innovation and the Bloomberg Philanthropies' Mayors Challenge, as well as being recognised as the number one smart city in the world by Juniper Research. He has published numerous essays and novels, for which he has received the Joan Fuster (2002), Enric Prat de la Riba (2003), Crexell (2011) and Llibreter (2014) prizes. His work reflects his extensive experience in the public and private sectors and his knowledge of city reality and its capacity to improve the lives of its inhabitants.

